Actuators

Scrivener Publishing
100 Cummings Center, Suite 541J
Beverly, MA 01915-6106

Publishers at Scrivener
Martin Scrivener (martin@scrivenerpublishing.com)
Phillip Carmical (pcarmical@scrivenerpublishing.com)

Actuators

Fundamentals, Principles, Materials and Applications

Edited by
Inamuddin, Rajender Boddula and Abdullah M. Asiri

Scrivener
Publishing

WILEY

Wiley Global Headquarters
111 River Street, Hoboken, NJ 07030, USA

For details of our global editorial offices, customer services, and more information about Wiley products visit us at www.wiley.com.

Limit of Liability/Disclaimer of Warranty
While the publisher and authors have used their best efforts in preparing this work, they make no representations or warranties with respect to the accuracy or completeness of the contents of this work and specifically disclaim all warranties, including without limitation any implied warranties of merchantability or fitness for a particular purpose. No warranty may be created or extended by sales representatives, written sales materials, or promotional statements for this work. The fact that an organization, website, or product is referred to in this work as a citation and/or potential source of further information does not mean that the publisher and authors endorse the information or services the organization, website, or product may provide or recommendations it may make. This work is sold with the understanding that the publisher is not engaged in rendering professional services. The advice and strategies contained herein may not be suitable for your situation. You should consult with a specialist where appropriate. Neither the publisher nor authors shall be liable for any loss of profit or any other commercial damages, including but not limited to special, incidental, consequential, or other damages. Further, readers should be aware that websites listed in this work may have changed or disappeared between when this work was written and when it is read.

Library of Congress Cataloging-in-Publication Data

ISBN 9781119661146

Cover image:
Cover design by Russell Richardson

Set in size of 11pt and Minion Pro by Manila Typesetting Company, Makati, Philippines

10 9 8 7 6 5 4 3 2 1

Contents

Preface

An actuator is a kind of part of a machine that is responsible for moving and controlling a mechanism/system. It can also be defined as something that converts energy into motion. The common types of actuators in automation include pneumatic, hydraulic, electromechanical, and mechanical actuators. Current technology has certain common needs for actuators designed for automotive, aeronautics, biomedical, robotics, and spatial applications. Today's technology aims towards designing of nano-, micro-, and macroscales for mechanical devices that can change their shape concerning the environmental conditions. The attention on developing actuating devices has been escalating in the past decade as a consequence of soaring demand for biomimetic multifaceted mechanisms such as implantable neuronal devices, animal-like robots, tissue substitutes, etc. In recent years, all over the world, researchers have concentrated on the development of new kinds of actuators owing to increasing demand for high-precision positioning technology in the areas of scientific and industrial research.

Actuators: Fundamentals, Principles, Materials and Applications aims to explore cutting-edge technology on actuators. The chapters discuss basics, principles, use of materials, types of actuators, and applications of actuators in mechatronics, robotics, artificial muscles, and high-precision positioning technology. It also includes actuators based on hydrogels, stimuli-responsive electroactive polymers, and smart-polymers. The challenges and prospects are also discussed. The book incorporates industrial applications and will fill the gap between the lab scale and practical applications. It will be of interest to engineers, industrialists, undergraduate and postgraduate students, faculty, and professionals. Based on thematic topics, the book contains the following nine chapters:

Chapter 1 discusses the piezoelectric actuators along with their applications in various fields. It is concluded that owing to the special properties, the piezoelectric actuators attracted much attention compared to other actuators, and research is going on to introduce new kinds of piezoelectric materials for actuator applications.

Chapter 2 reveals some of the important parameters that are to be considered while designing a shape memory alloy (SMA) actuated system. The design parameters will enable us to build an efficient control mechanism and also give a broad range of basic elements that are to be decided before the developing stage, and the parameters required are quantified.

Chapter 3 captures actuators in the mechatronics system or robotics. This chapter discusses in detail the various types of actuators such as pneumatic, hydraulic, mechanical, and electromechanical. The various components of actuators, as well as applications of actuators, are also stated.

Chapter 4 discusses the use of stimuli-responsive hydrogels for the development of soft-actuators with main applications in automated biomedical devices due to its tissue-equivalence. Important concepts about the properties, synthesis, and characterization of hydrogels and their respective soft-actuators are presented. This chapter can help as a practical guide to consulting current fundamental reactions and manufacturing methodologies.

Chapter 5 reviews various polymer-based chemical sensors for the detection of different parameters such as gas, vapors, humidity, pH, and ions. The exploration of numerous kinds of polymers, along with the utilization of different techniques with their merits and demerits, is also elaborately discussed. This chapter also highlights the future perspective of polymer-based chemical sensors.

Chapter 6 deals with various shape memory actuators, which mainly include shape memory alloy (SMA) actuators and shape memory polymer (SMP) actuators. The classification of shape memory effects (SME), types of actuators, advantages, applications, and various design mechanisms, or strategies of these two shape memory materials (SMM) is also discussed in detail.

Chapter 7 reports the current progress of stimuli-responding conducting polymer (CP) composites to understand the mechanical behavior of these materials concerning electrical-, photo-, and thermo-responsive stimuli. An overview of the most widely used CPs composite materials with their uses and operational mechanisms, as well as a specified account of CPs as next-generation actuators, is also presented.

Chapter 8 discusses the various types of fluid power actuators like single and double acting cylinders, telescopic cylinder and tandem cylinders used for industrial applications. This chapter also discusses the application of fluid power actuators for robotics, legged robotics and MEMS application.

Chapter 9 provides an overview of conducting polymer/hydrogel systems as soft actuators, conducting polymer actuators and their actuation mechanism, their merits and demerits followed by highlighting the progress of CP/hydrogel actuators fabricated using polypyrrole, polyaniline and polythiophene with various hydrogels, and outlines the factors affecting their actuation performance.

Key Features

1. A comprehensive and self-contained text covering all aspects of the multidisciplinary fields of actuators
2. Contains contributions from noted experts in the field
3. Provides a broad overview of actuators
4. Deliberates cutting-edge technology based on actuators

Editors
Inamuddin
Rajender Boddula
Abdullah M. Asiri

1

Piezoelectric Actuators and Their Applications

N. Suresh Kumar[1]*, R. Padma Suvarna[1], K. Chandra Babu Naidu[2†],
S. Ramesh[2], M.S.S.R.K.N. Sarma[2], H. Manjunatha[3],
Ramyakrishna Pothu[4] and Rajender Boddula[5]

[1]Department of Physics, JNTUA, Anantapuramu, India
[2]Department of Physics, GITAM Deemed to be University, Bangalore, India
[3]Department of Chemistry, GITAM Deemed to be University, Bangalore, India
*[4]College of Chemistry and Chemical Engineering, Hunan University,
Changsha, China*
*[5]CAS Center for Excellence in Nanoscience, National Center for Nanoscience
and Technology, Beijing, China*

Abstract

Piezoelectric actuators (PEAs) are a type of microactuators which mainly use the inverse piezoelectric effect to produce small displacement at high speed by applying voltage. This chapter includes the detailed discussion on piezoelectric actuators in the direction of industrial benefits. In addition, the classification of piezoelectric actuators is made. Especially, the piezoelectric actuators showed the MEMS (microelectromechanical systems) applications at larger extent. We elaborated different piezoelectric materials such as lead and lead zirconium-based compounds for actuator applications. In view of this, few parameters like memory, domain rotation, etc., are considered for justifying the actuator applications. The importance of these actuators towards robotics is also elucidated.

Keywords: Actuators, piezoelectric materials, hydraulic actuator, journal bearings, machines

Corresponding author: drsureshkumarnagasamudram@gmail.com
†*Corresponding author*: chandrababu954@gmail.com

Inamuddin, Rajender Boddula and Abdullah M. Asiri (eds.) Actuators: Fundamentals, Principles,
Materials and Applications, (1–16) © 2020 Scrivener Publishing LLC

1.1 Introduction

A part of a device which moves or controls the mechanism is called an actuator. Example is an electric motor which converts a control signal to mechanical action. An actuator is one which converts energy into motion. This energy may be hydraulic, pneumatic, electric, thermal, mechanical, or even human power [1]. Whenever a control signal is received, actuator converts the energy of the signal into mechanical motion. The control system may be a fixed mechanical/electronic system or a software-based printer device/robot. Based on the type of control system or energy, actuators are classified into different types which are hydraulic, pneumatic, electrical, mechanical actuators, etc.

1.2 Types of Actuators

A hydraulic actuator uses hydraulic power for the mechanical process. Here, the output will be a linear, rotary, or oscillatory motion. In hydraulic systems, energy is transmitted with the help of pressure of fluid in a sealed system. It has advantages like efficient power transmission, accuracy, and also flexible in maintenance. Usage of these systems is safe in chemical plants and mines as they do not produce any sparks. Here, leakage of the fluid is the major drawback. Because, once the fluid leaks, it may catch fire or leads to serious injuries when it bursts [2]. Examples are brakes in cars/trucks, wheelchair lifts, hydraulic jacks, and flaps on aircrafts, etc. Figure 1.1 represents one type of hydraulic actuator.

A pneumatic actuator rotation is shown in Figure 1.2. It uses energy in the form of compressed air at high pressure to produce motion. This actuator is mainly advisable in main engine controls for swift starting and stopping. These actuators are economical, lightweight, less maintenance, and simple when compared to other actuators. The disadvantage with this actuator is application-specific that is an actuator sized for a specific purpose cannot be used for other applications [3, 4].

An electric actuator uses a motor for converting electrical energy into mechanical torque. This is used in multi-turn valves. Figure 1.3 shows different types of electric actuators. As there is no involvement of either fossil fuels or oils these actuators are the cleanest and easily available ones [5, 6]. Another type of actuator is supercoiled polymer (SCP) or twisted and coiled polymer (TCP) actuators which use electric power for actuating. These are made up of silver-coated nylon and gold and appear helical like a

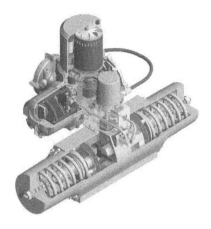

Figure 1.1 Hydraulic actuator.

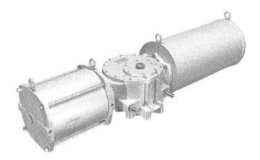

Figure 1.2 Pneumatic actuator.

Figure 1.3 Electric actuators.

splicing. These are constructed by twisting like nylon thread such as fishing line. They serve as bicep muscle to control the motion of arms in robots. Because of electrical resistance, electrical energy gets converted into thermal energy also called Joule heating. When the temperature of this actuator is increased due to joule heating, counteraction of the polymer takes place resulting in contraction of the actuator [7].

Mechanical actuator converts one type of motion into another like rotatory into linear. Example for this type is rack and pinion. This type of actuators depends on constitutional components like gears and rails, pulleys, and chairs. In order to use actuators in the fields of agriculture for fruit harvesting and biomedicine in robotics soft actuators are being developed. Because of amalgamation of microscopic changes at the basic (molecular) level into a macroscopic distortion, soft actuators generate flexible motion. Figure 1.4 represents mechanical actuator.

Electromechanical actuators are nothing but mechanical actuators in which the control signal is given by an electric motor which converts the rotatory motion into linear displacement. These actuators work on the inclined plane concept. The required inclination is provided by the threads of a lead screw. It acts as a ramp and converts small rotational force into linear displacement. Based on parameters like operation, speed increased load capacity, mechanical efficiency, etc., different EM actuators are designed which are shown in Figure 1.5. The common design consists of a lead screw passing through the motor. The lead nut is the only moving part while the lead screw remains fixed and non-rotating. When the motor rotates, the lead screw either extends outwards or retracts inwards. By using alternating threads on the same shaft, different actuators are designed. Actuators begin on the lead screw and provide a higher adjustment capability between the starts and the nut thread area of contact, influencing the extension speed and load capacity.

The density of motion of the nut is determined by the lead screw and by coupling the linkages to the nut. Usually, in many cases, screw is connected to the motor. Based on the amount of the loads, the actuator is expected to move various motors like dc brush, stepper, induction motors, etc. Coming to its advantages, the person handling this actuator can have complete control over the movement. They can control the velocity and position accurately without switching off the device or when the device is in running state, the force and motion profile can be changed by changing its software. These actuators consume power only when they are in operation. Low maintenance, high efficiency, and being environmentally friendly make these actuators a potential candidate in hazardous areas.

Figure 1.4 Mechanical actuator.

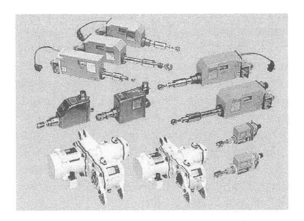

Figure 1.5 Electromechanical actuators.

These are used in packaging, food, energy process control, construction, and automation industry.

1.3   Piezoelectric Actuators

In recent years, all over the world, the researchers concentrated on the development of new kinds of precision actuators owing to increase in demand for high precision positioning technology in the areas of scientific and industrial research [8–15]. In specific, piezoelectric actuators (PEAs) are gaining much attention due to their novel properties like fast response, compact structure, high precision, etc. PEAs are a type of microactuators which mainly uses the inverse piezoelectric effect to produce small displacement at high speed by applying voltage [16–20]. In recent years, these PEAs have enhanced their area of applications, upgraded synthesis techniques [21], and also improved properties have led to different types of designs which can accomplish large displacements with reasonable voltages even though maintaining comparatively large stiffness [22, 23]. In this chapter, we discussed different types of PEA and their applications in various fields.

Adriaens *et al.* [24] introduced the electromechanical model for PEAs. They improved the usage of nonlinear first-order differential equation to explain the effect of hysteresis and the use of structural damping and partial differential equation (PDE) to explain the mechanical behavior. Furthermore, they concluded that the hysteresis effect and disseminated nature of PEA can be circumvented through proper design of the positioning mechanism and also varying traditional voltage steering for charge steering. Hence, the simplified mechanism (piezo-actuated) is very much appropriate as a controller design basis.

Cattafesta *et al.* [25] discussed the development of piezoelectric unimorph flap actuators for active flow control. They designed this model with the help of composite beam model (CBM) along with optimization method. Even though, CBM cannot capture the total information about the behavior like anticlastic curvature of the actuator. Anyhow, this type of simple model is suitable for understanding the relationship between the design variables and performance of the actuator. Nevertheless, to study the complex geometries, a finite element or analytical plate model is much appropriate. They reported that PEA is designed by using optimized model. In addition, they also discussed a simple application, which is the controlling of flow from a backward-facing step where the actuator is fixed at origin of the shear layer. Fig. 4 in reference [25] shows the actuator installation in back word-facing step. Furthermore, we know that the fluid-structure model assumed that the displacement of the actuator is proportional to the induced streamwise speed trepidation in the flow. Quasi-static model also describes that, due to quasi-static motion of the actuator, the shear layer effectually moves in vertical direction. Finally, they reported that the developed model leads to a proportionate correlation between the incoming boundary layer, streamwise speed variations induced by the actuator and the displacement of the flap tip which is very much suitable for the designing actuators for flap control applications.

Based on the theory of elasticity, Shi [26] investigated and compared the bending behaviors of a functional gradient curved piezoelectric actuator (FGPEA) and a bi-morph curved piezoelectric actuator (BPEA). He reported that the exact solutions of both electrical and mechanical fields are attained. Besides, when the curved FGPEA is subjected to an external electric filed, there exist non-zero internal stresses whereas in flat FGPEA no stress is observed. However, the deflection of the FGPEA is somewhat lesser than that of BPEA but the considerable reduction in internal stresses is noticed in FGPEA. Taotao Zhang and Zhifei Shi [27] reported the similar kind of outcomes in FGPEA and piezoelectric curved multi-layered actuators.

Kadota and Morita [28] fabricated a unimorph shape memory piezoelectric actuator (SMPEA) by using a soft PZT. A SMPEA is an actuator which can exhibit a piezoelectric displacement without an external voltage. These actuators have two stable states at zero voltage which are poled state and depoled state. Mainly, the driving principle in SMPEAs depends on reorientation of non 180° domains. They reported that the investigations on the prepared SMPEAs revealed that the fatigue properties were studied by continuous switching of cycling polarization, owing to reorientation of non 180° domains and the prepared SMPEAs exhibited superior properties when compared to existing actuators [29, 30]. During early stages, the fatigue is caused by domain pinning while in later stages it is caused by damaged layer. However, the observed shape memory displacement is not constant, which varies as function of time.

Yoshida *et al.* [31] prepared a high AR PZT (aspect ratio lead zirconate titanate) structures via nanocomposite sol-gel technique. Further, the prepared material is sandwiched between two Pt electrodes (prepared by ALD and photolithographic techniques) to fabricate an actuator for MEMS (microelectromechanical systems) applications. They successfully presented the demonstrations of both unimorph and bimorph actuation. The outcomes revealed that the d_{33} coefficient about 20 pC N^{-1} and a 10-µm lateral displacement is observed for 500-µm actuator in bimorph actuation. They reported that the developed actuator occupies small area and generates similar displacement compared to conventional electrostatic combo-drive actuators; therefore, it can be for diverse appliances in MEMS.

Li and Chen [32] introduced a new hysteresis model for describing the behavior of hysteresis in PEA accurately. They reported that the proposed model is an online adaptive rate-dependent hysteresis model which exhibits high efficiency and precision when compared with the offline model. This new model shows excellent rate-dependent property which is in well agreement with the actual hysteresis behavior in PEA. In addition, the experimental outcomes revealed that the hysteresis in PEA can be evidently controlled and allows observing the position and performance accuracy of the order of 10-nm range. One more advantage of the proposed model is it can track the non-periodic random movement. From the results of the experiment, it is evident that the proposed model using PEA can be efficiently useful for attaining ultra-precession of the motion control.

Jiang *et al.* [33] proposed a novel hybrid photovoltaic/piezoelectric actuation mechanism. They reported that PbLaZrTi (PLZT) ceramic materials are utilized as a photovoltaic generator to drive PEA. The fruitful model of this new mechanism of actuation is presented and verified experimentally. Some significant characteristics of the newly developed model are analyzed and reported that there exists no hysteresis between deformation which exists in PEA and applied voltage. With the help of multi-patch PbLaZrTi generator, one can improve the response speed and saturated voltage to drive the PEA, and by using high light intensity, the initial voltages which arise through a logical switch between the PEA and PLZT can be improved. Hence, this new model can be useful as control at high frequencies.

Orszulik and Shan [34] reported an output feedback integral control of the PEA by considering the effect of hysteresis. In the outcome, they described that, in the proposed model, the nonlinear part represents the hysteresis with respect to linear part. In addition, the Lipschitz condition is derived by considering the Prandtl-Ishlinskii model as the hysteresis. With respect to frequency response, the linear parameters of the proposed model are obtained by using experimental single-axis PEA while the hysteresis model parameters are attained by using the procedure of nonlinear least squares. Finally, they proposed that the derived maximum integral gain of the PEA drives the experimental system which reveals that the system with closed loop is certainly stable.

Continuing the development of piezoelectric actuators, Patil *et al.* [35] introduced a new type of piezoelectric actuator and sensor which is integrated into the structures to detect the bolt loosening in that frame structures. By considering the dynamic responses from undamaged and damaged structures, the location of damage can be observed and estimated with the help of FRF (frequency response function) method together with PEA. From the outcomes of the simulation work (carried out with the help of MATLAB and ABAQUS) shown in Table 5 of the reference [35], it is evident that the PEA can be effectively used to find the damages caused by the bolt removal.

Tuma *et al.* [36] reported that how the characteristics performance of the journal bearing [JB] is affected by the active vibration control (AVC) by piezo-actuators. Before publishing this article, many researchers have shown interest to the active control of the JBs by using GMM (giant magnetostrictive material) [37] and AMB (active magnetic bearings) [38]. From the experiments carried with GMM, Tuma *et al.* [36] stated that the problem arises at high rpm of around 1,700 owing to instability of the JB. Also, they published that, at low rpm of about 350, the instability did not arise. Hence, AVC is not intended at eradicating the instability of JBs but only on placing the shaft axis. The development of piezoelectric actuators has been started from early 1990s [39] which can be used to control the rotating machines. In addition, the linear piezoelectric actuators can produce large amount of force in a small track. The main advantage of the piezoelectric actuators compared to AMB is unchanged bearing bushing mounting stiffness in the journal bearings. Fig. 2 and Fig. 3 in the reference [36] shows the actively controlled journal bearing and test bench. Finally, they concluded that the usage of piezoelectric actuator in journal bearing causes the increase in stable operating range, losses due to friction is reduced and radial stiffness is increased.

Long *et al.* [40] implemented a novel synthesis technique, i.e., combining the conventional SSR (solid-state reaction) with gel-precursors mixing technique to prepare lead-free BCTS $(Ba_{1-x}Ca_x)$ $(Ti_{0.91}Sn_{0.09})O_3)$ ceramic materials. They reported that the prepared ceramic materials exhibit high performance with uniformly distributed grain size and boundary. In addition, the electrical properties are augmented, especially for x = 0.03 compound the value of d_{33} coefficient is 620 pC/N, dielectric constant is 14,500 and high dielectric tunability (>86.97%) are observed. Besides, the prepared lead-free ceramic actuators exhibit large displacement at low operating voltages which indicates that the prepared ceramics have potential appliances in multilayer piezoelectric devices.

Li *et al.* [41] investigated the micro-displacement mechanism (electrostrictive and inverse piezoelectric effect) of the PEAs and also the reversal of electric domain due to ferroelectric effect. From the results, it is clear that the displacement mainly arises due to ferroelectric and inverse piezoelectric effect, whereas the electrostrictive is almost ignored (extremely weak) at microlevel. They also observed that there is a linear relationship between the displacement and voltage in the PEA in inverse piezoelectric effect and hysteresis occurs in ferroelectric effect. In addition, some irreversible non-180° domains are the basis of the hysteresis of the PEA. The proposed model improves the accuracy of PEAs in practical applications.

Recently, owing to increased usage of mechatronic systems, researchers concentrated on designing of highly automated systems. FDI (fault detection and isolation) is a key concern to confirm their operations. Continuing this, Ramakrishnan Ambur and Stephan Rinderknecht [42] designed a parameter estimation method in order to detect the unbalanced faults (phase, magnitude, and location) in a rotating machine with PEA. These actuators can also be acted as sensors due to intrinsic connection between their mechanical and electrical properties. From these self-sensing actuators, reconstruction of mechanical deflection takes place and also these are used to find unbalance in rotor system. The WLS (weighted least squares) method is used to improve the robustness of the estimations against outliers in measurement. Fig. 11 in reference [42] exhibits the detection of unbalances in real test bench. They reported that all the unbalanced faults are successfully detected in case of stationary tests while in case of closed-loop and transient runs test only the magnitude is detected.

Ciou and Lee [43] designed a controllable preload spindle mechanism for the tools in machines and tested using a single PEA. Fig. 1 in reference [43] shows the wedge-shaped spacer ring which is enabled with the help of a PEA to produce uniform axial displacement to vary the preload on bearings. The relationship between the stiffness and displacement of the spindle is evaluated by using load-displacement model. From the results, it is clear that the increased preload causes the reduction in vibrational amplitude and increase in temperature of the bearing. So, the designed model is inexpensive and viable approach for controlling preload spindles.

Alireza *et al.* [44] investigated on nonlinear dynamics and statics of a slightly curved Euler-Bernoulli capacitive microbeam (bi-stable microelectromechanical system) enclosed between two piezoelectric layers. They reported that Hamilton's principle is used to derive the governing equation of motion. In addition, the size effect is formulated via strain gradient theory which reduced to set of differential equations depending on Galerkin approach. Moreover, the proposed model is responsible for nonlinearities owing to curvature, electrostatic loading and stretching at midplane. Owing to the application of DC voltage on the piezoelectric layers, the change in static bi-stability band of the system takes place and also the pull in and snap-through instabilities can be ordered. Furthermore, primary resonance (Fig. 13 in reference [44] exhibits the influence of piezoelectric actuation) is categorized with the help of the shooting method and Floquet theorem is used to describe limit cycles stability. Finally, the dynamic snap-through band could be changed effectively by applying proper piezoelectric actuation. The designed model can be applied in designing micro arch-based bandpass filters, smartwatches, etc.

Deng *et al.* [45] proposed a PEA operating at two actuating modes (direct and inertial actuation) and tested. They reported that the proposed PEA exhibits sandwich structure with large resonant frequency which is different from other PEAs (flexure structures). In addition, using scanning laser Doppler vibrometer, the resonance frequency is observed to be 12,720 Hz. Besides, when the PEA is operated at direct mode, DC signal is applied and attained a displacement resolution of 1.25 cm with a highest scan-rate and range of around 190 Hz and 3685 nm. Furthermore, in inertial mode of actuation, there is linear relation among the voltage and speed which may be used as linear dynamic control of the speed. Here, the achieved maximum speed is around 14.439×10^{-6} m/s and thrust force is 1.66 N. Hence, these types of sandwich structured piezoelectric actuators provide a linear driving with extended stroke (inertial actuation mode) and maximum scan rate and nanometer resolution (direct actuation mode) which have proper application possibilities in the areas of biological manipulation and nano-scanning.

Lallart *et al.* [46] proposed a simple model for predicting the quasi-static operations, i.e., the relationship between strain and voltage in PEAs. Fig. 6 in reference [46] shows the block diagram of the proposed model. They reported that in the proposed model, an electrochemical coupling (butterfly-shaped) lies in between the strain and voltage derivatives. In addition, this model works in both unipolar and symmetric cycles. The experimental outcomes exhibited that the unipolar PEA is well matched with the theoretical values. In general, the proposed model can also be applied for other domains like hydraulics, magnetic actuation magnetism, etc.

Lu *et al.* [47] investigated a sensorless (self-sensing) control of MPEA (multilayer piezoelectric actuator) displacements, particularly for low-speed operations (below 100 Hz); however, for high-speed operations, the proposed model monitors whether the operation is at resonance or not.

Ye *et al.* [48] designed a novel joint-arm (ring-beam) PEA for manipulator, which works on the principle of piezoelectric excitation and with the help of converse piezoelectric effect the actuator is driven directly by the frictional force induced at joint interface. They reported that the experimental outcomes revealed the response time (start up and shut down) of the developed actuator is very small, of the order of 31 ms and 21 ms. In addition, the angular speed of the model reaches 45.90 deg/sec with a resolution of

0.015 deg. From the obtained results, it is clear that, due to high resolution and fast response rate, proposed actuator can be utilized to develop a manipulator with high precision and good mobility. Wang et al. [49] proposed a new type piezoelectric inertial rotatory motor which works on the principle of slip-slip mechanism for actuating the MUVs (micro underwater vehicles). Fig. 16 and Fig. 17 in reference [49] exhibit the experimental outcomes of the proposed motor. Masoud Soltan Rezae et al. [50] introduced a nonlinear model by regulating piezoelectric microwires with the capacity to adjust the stability conditions to enhance characteristics of the system. The appropriate adjustments are instigated electrostatically by actuating or applying a small piezo voltage to enhance the properties of the system such as tuning applications, controllability, etc. They reported that the experimental outcomes of the present model are helpful in the development of new advanced actuated microdevices. Tian et al. [51] designed a novel U-shaped stepping PEA containing low motion coupling. In this model, the actuator contains 4-PAUs (piezoelectric actuation units); among them, two acts as clamping and remaining two acts as pushing units. Fig. 3 in reference [51] shows the operating principle of the proposed actuator. They reported that the proposed actuator can be employed in the areas of precision zoom lens adjustment system, precision positioning platform for microscope, precision feeding of machine tool, etc.

In general, charge measurement technique is used to estimate the position of PEAs. In this, linear charge position property is helpful in finding the actuator position. However, charge-position relation is non-linear in low impedance actuators besides internal resistance of the piezoelectric material induces some leakage of charge. Due to this, the traditional charge-based approaches are not applicable for low impedance actuators. To overcome these problems, Soleymanzadeh et al. [52] introduced a new model in which they considered the leakage of charge and other uncertainties such as dielectric absorption, current biases, etc., due to sensor fault. In the proposed method by employing the FDI (fault detection and isolation) based robust observer, the position of the actuator is estimated properly and experimental results showed the efficiency of the proposed model for the estimation of position in multi-frequency and quasi-static cases with an error of 0.1453 μm and 0.1907 μm. Therefore, the proposed model may be applicable for low impedance actuators. In this chapter, a modest attempt has been made on the piezoelectric actuators from this; it is clear that due to their special properties, the piezoelectric actuators attracted much attention compare to other actuators.

1.4   Conclusions

We discussed about the piezoelectric actuators along with their applications in various fields. It was concluded that, owing to the special properties, the piezoelectric actuators attracted much attention compare to other actuators and research is going on to introduce new kinds piezoelectric materials for actuator applications.

References

1. About Actuators. www.thomasnet.com. Archived from the original on 2016-05-08. Retrieved 2016-04-26.
2. What's the Difference Between Pneumatic, Hydraulic, and Electrical Actuators? machinedesign.com. Archived from the original on 2016-04-23. Retrieved 2016-04-26.
3. What is a Pneumatic Actuator?. www.tech-faq.com. Archived from the original on 2018-02-21. Retrieved 2018-02-20.
4. Pneumatic Valve Actuators Information/IHS Engineering 360. www.global-spec.com. Archived from the original on 2016-06-24. Retrieved 2016-04-26.
5. Tisserand, O., "How does an electric actuator work?". Archived from the original on 2018-02-21. Retrieved 2018-02-20.
6. Electric & Pneumatic Actuators. www.baelzna.com. Archived from the original on 2016-04-30. Retrieved 2016-04-26.
7. Jafarzadeh, M., Gans, N., Tadesse, Y., Control of TCP muscles using Takagi–Sugeno–Kang fuzzy inference system. *Mechatronics.*, 53, 124–139, 2018. https://doi.org/10.1016/j.mechatronics.2018.06.007
8. Chen, M.Y., Tsai, C.F., Fu, L.C., A novel design and control to improve positioning precision and robustness for a planar maglev system. *IEEE Trans. Ind. Electron.*, 66, 4860–4869, 2019. https://doi.org/10.1109/TIE.2018.2821633
9. Gao, W., Kim, S.W., Bosse, H., Haitjema, H., Chena, Y.L., Lu, X.D., Knapp, W., Weckenmann, A., Estler, W.T., Kunzmann, H., Measurement technologies for precision positioning. *CIRP Ann.-Manuf. Technol.*, 64, 773–796, 2015. https://doi.org/10.1016/j.cirp.2015.05.009
10. Ling, M., Cao, J., Jiang, Z., Lin, J., Modular kinematics and statics modeling for precision positioning stage. *Mech. Mach. Theory*, 107, 274–282, 2017. https://doi.org/10.1016/j.mechmachtheory.2016.10.009
11. Xu, W.X. and Wu, Y.B., Piezoelectric actuator for machining on macro-to-micro cylindrical components by a precision rotary motion control. *Mech. Syst. Signal. Process.*, 114, 439–447, 2019. DOI:10.1016/j.ymssp.2018.05.035
12. Gu, G.-Y., Zhu, L.-M., Su, C.-Y., High-precision control of piezoelectric nano positioning stages using hysteresis compensator and disturbance

observer. *Smart Mater. Struct.*, 23, 105007, 2014. https://doi.org/10. 1088/0964-1726/23/10/105007

13. Yu, C.F., Wang, C.L., Deng, H.S., He, T., Mao, P.F., Hysteresis nonlinearity modeling and position control for a precision positioning stage based on a giant Magnetostrictive actuator. *RSC Adv.*, 6, 59468–59476, 2016. DOI:10.1039/C6RA05195B

14. Ito, S. and Schitter, G., Comparison and classification of high-precision actuators based on stiffness influencing vibration isolation. *IEEE/ ASME Trans. Mechatron.*, 21, 1169–1178, 2016. https://doi.org/10.1109/ TMECH.2015.2478658

15. Petit, L., Dupont, E., Lamarque, F., Prelle, C., Design and characterization of a high-precision digital electromagnetic actuator with four discrete positions. *Actuators*, 4, 217–236, 2015. doi:10.3390/act4040217

16. Morita, T., Miniature piezoelectric motors. *Sens. Actuators A, Phys.*, 103, 291–300, 2003. https://doi.org/10.1016/S0924-4247(02)00405-3

17. Huang, S., Tan, K.K., Lee, T.H., Adaptive sliding-mode control of piezoelectric actuators. *IEEE Trans. Ind. Electron.*, 56, 3514–3522, 2009. DOI: 10.1109/ TIE.2009.2012450

18. Liu, J.K., Liu, Y.X., Zhao, L.L., Xu, D.M., Chen, W.S., Deng, J., Design and experiments of a single-foot linear piezoelectric actuator operated in stepping mode. *IEEE Trans. Ind. Electron.*, 65, 8063–8071, 2018. https://doi. org/10.1109/TIE.2018.2798627

19. Chen, W.S., Liu, Y.X., Yang, X.H., Liu, J.K., Ring type traveling wave ultrasonic motor using radial bending mode. *IEEE Trans. Ultrason. Ferroelectr. Freq. Control*, 61, 197–202, 2014. DOI: 10.1109/TUFFC.2014.6689788

20. Liu, Y.X., Yang, X.H., Chen, W.S., Xu, D.M., A bonded-type piezoelectric actuator using the first and second bending vibration modes. *IEEE Trans. Ind. Electron.*, 63, 1676–1683, 2016. DOI: 10.1109/TIE.2015.2492942

21. Uchino, K., *Advanced Piezoelectric Materials: Science and Technology*, p. 9, Woodhead Publishing, Cambridge, 2010.

22. Watson, B., Friend, J., Yeo, L., Modelling and testing of a piezoelectric ultrasonic micro-motor suitable for *in vivo* micro-robotic applications. *J. Micromech. Microeng.*, 20, 115018, 2010. https://doi.org/10.1088/ 0960-1317/20/11/115018

23. De Lorenzo, D., De Momi, E., Dyagilev, I., Manganelli, R., Formaglio, A., Prattichizzo, D., Shoham, M., Ferrigno, G., Force feedback in a piezoelectric linear actuator for neurosurgery. *Int. J. Med. Robot. Comput. Assist. Surg.*, 7, 268–275, 2011. doi: 10.1002/rcs.391. Epub 2011 Apr 28.

24. Adriaens, H. J. M. T. A., de Koning, W.L., Banning, R., Modeling Piezoelectric Actuators. *IEEE/ASME Trans. Mechatron.*, 5, 331–341, 2000. DOI: 10.1109/3516.891044

25. Cattafesta, L.N., Garg, S., Shukla, D., Development of Piezoelectric Actuators for Active Flow Control. *AIAA J.*, 39, 1562–1568, 2001. https://doi. org/10.2514/2.1481

26. Shi, Z., Bending behavior of piezoelectric curved actuator. *Smart Mater. Struct.*, 14, 835–842, 2005. https://doi.org/10.1088/0964-1726/14/4/043

27. Zhang, T. and Shi, Z., Two-dimensional exact analysis for piezoelectric curved actuators. *J. Micromech. Microeng.*, 16, 640–647, 2006. https://doi.org/10.1088/0960-1317/16/3/020

28. Kadota, Y. and Morita, T., Fatigue and retention properties of shape memory piezoelectric actuator with non-180° domain switching. *Smart Mater. Struct.*, 21, 045002 (10pp), 2012. https://doi.org/10.1088/0964-1726/21/4/045002

29. Kadota, Y., Hosaka, H., Morita, T., Shape memory piezoelectric actuator by control of the imprint electrical field. *J. Ferroelectr.*, 368, 185–193, 2008. https://doi.org/10.1080/00150190802368479

30. Kadota, Y., Hosaka, H., Morita, T., Fatigue property of shape memory piezoelectric actuator by continuous cycling of polarization switching. *Mater. Res. Soc. Symp. Proc.*, 1110E, 1110-C09-13, 2009. DOI: 10.1557/PROC-1110-C09-32

31. Yoshida, S., Wang, N., Kumano, M., Kawai, Y., Tanaka, S., Esashi, M., Fabrication and characterization of laterally-driven piezoelectric bimorph MEMS actuator with sol–gel-based high-aspect-ratio PZT structure. *J. Micromech. Microeng.*, 23, 065014 (11pp), 2013. https://doi.org/10.1088/0960-1317/23/6/065014

32. Li, W. and Chen, X., Compensation of hysteresis in piezoelectric actuators without dynamics modeling. *Sens. Actuators A*, 199, 89–97, 2013. https://doi.org/10.1016/j.sna.2013.04.036

33. Jiang, J., Li, X., Ding, J., Yue, H., Deng, Z., Mathematical model and characteristic analysis of hybrid photovoltaic/piezoelectric actuation mechanism. *Smart Mater. Struct.*, 25, 125021 (10pp), 2016. https://doi.org/10.1088/0964-1726/25/12/125021

34. Orszulik, R.R. and Shan, J., Output Feedback Integral Control of Piezoelectric Actuators considering Hysteresis. *Precis. Eng.*, 47, 90–96, 2017. https://doi.org/10.1016/j.precisioneng.2016.07.009

35. Patil, C.S., Roy, S., Jagtap, K.R., Damage Detection in Frame Structure Using Piezoelectric Actuator. *Mater. Today: Proc.*, 4, 687–692, 2017. https://doi.org/10.1016/j.matpr.2017.01.073

36. Tuma, J., Šimek, J., Mahdal, M., Pawlenka and R Wagnerova, M., Piezoelectric actuators in the active vibration control system of journal bearings. *IOP Conf. Ser.: J. Phys.: Conf. Ser.*, 870, 012017, 2017. https://doi.org/10.1088/1742-6596/870/1/012017

37. Lau, H.Y., Liu, K.P., Wang, W., Wong, P.L., Feasibility of Using Giant Magnetostrictive Material (GMM) Based Actuators in Active Control of Journal Bearing System, in: *Proceedings of the World Congress on Engineering*, vol. II, WCE, London, U.K, 2009, July 1 - 3, 2009.

38. Fürst, S. and Ulbrich, H., An Active Support System for Rotors with Oil-Film Bearings, in: *Proceedings of IMechE, Serie C*, pp. 61–68, paper 261/88, 1988.

39. Palazzolo, A.B., Lin, R.R., Alexande, R.M., Kascak, A.F., Montague, G., Test and Theory of Piezoactuators - Active Vibration Control of Rotating Machinery, ASME Trans. *J. Vib. Acoust.*, 113, 167–175, 1991. DOI: 10.1115/1.2930165

40. Long, P., Liu, X., Long, X., Yi, Z., Dielectric relaxation, impedance spectra, piezoelectric properties of (Ba, Ca) (Ti, Sn)O$_3$ ceramics and their multilayer piezoelectric actuators. *J. Alloys Compd.*, 706, 234–243, 2017. https://doi. org/10.1016/j.jallcom.2017.02.237

41. Li, H., Xu, Y., Shao, M., Guo and D An, L., Analysis for hysteresis of piezoelectric actuator based on microscopic mechanism. *IOP Conf. Ser.: Mater. Sci. Eng.*, 399, 012031, 2018. https://doi.org/10.1088/1757-899X/399/1/012031

42. Ambur, R. and Rinderknecht, S., Unbalance detection in rotor systems with active bearings using self-sensing piezoelectric actuators. *Mech. Syst. Sig. Process.*, 102, 72–86, 2018. https://doi.org/10.1016/j.ymssp.2017.09.006

43. Ciou, Y.S. and Lee, C.Y., Controllable preload spindle with a piezoelectric actuator for machine tools. *Int. J. Mach. Tools Manuf.*, 102, 72–86, 2018. https://doi.org/10.1016/j.ijmachtools.2019.01.004

44. Nikpourian, A., Ghazavi, M.R., Azizi, S., Size-dependent nonlinear behavior of a piezoelectrically actuated capacitive bistable microstructure. *Int. J. Non-Linear Mech.*, 114, 49–61, 2019. https://doi.org/10.1016/j.ijnonlinmec.2019.04.010

45. Deng, J., Chen, W., Li, K., Wang, L., Liu, Y.X., A sandwich piezoelectric actuator with long stroke and nanometer resolution by the hybrid of two actuation modes. *Sens. Actuators A*, 296, 121–131, 2019. https://doi.org/10.1016/j.sna.2019.07.013

46. Lallart, M., Li, K., Yang, Z., Wang, W., System-level modeling of nonlinear hysteretic piezoelectric actuators in quasi-static operations. *Mech. Syst. Sig. Process.*, 116, 985–996, 2019. https://doi.org/10.1016/j.ymssp.2018.07.025

47. Lu, T.F., Fan, Y., Morita, T., An Investigation of Piezoelectric Actuator High Speed Operation for Self-sensing. *Measurement*, 136, 105–115, 2019. https:// doi.org/10.1016/j.measurement.2018.12.055

48. Ye, Z., Zhou, C., Jin, J., Yu, P., Wang, F., A novel ring-beam piezoelectric actuator for smallsize and high-precision manipulator. *Ultrasonics*, 96, 90–95, 2019. doi: https://doi.org/10.1016/j.ultras.2019.02.007. https://doi. org/10.1016/j.ultras.2019.02.007

49. Wang, L., Hou, Y., Zhao, K., Shen, H., Wang, Z., Zhao, C., Lu, X., A novel piezoelectric inertial rotary motor for actuating micro underwater vehicles. *Sens. Actuators A*, 295, 428–438, 2019. https://doi.org/10.1016/j.sna.2019.06.014

50. Soltan Rezaee, M., Bodaghi, M., Farrokhabadi, A., Hedayati, R., Nonlinear stability analysis of piecewise actuated piezoelectric microstructures. *Int. J. Mech. Sci.*, 160, 200–208, 2019. https://doi.org/10.1016/j.ijmecsci.2019.06.030

51. Tian, X., Zhang, B., Liu, Y., Chen, S., Yu, H., A novel U-shaped stepping linear piezoelectric actuator with two driving feet and low motion coupling: Design, modeling and experiments. *Mech. Syst. Sig. Process.*, 124, 679–695, 2019. https://doi.org/10.1016/j.ymssp.2019.02.019

52. Soleymanzadeh, D., Ghafarirad, H., Zareinejad, M., Charge-Based Robust Position Estimation for Low Impedance Piezoelectric Actuators. *Measurement*, 147, 106839, 2019. https://doi.org/10.1016/j.measurement.2019.07.067

Design Considerations for Shape Memory Alloy-Based Control Applications

Josephine Selvarani Ruth D* and Glory Rebekah Selvamani D

Robert Bosch Centre for Cyber Physical Systems, Indian Institute of Science, Bangalore, India

Abstract

Shape memory alloy (SMA) is one of the classes of smart material with high power to weight ratio, small size, and silent actuation. This has an ability of external actuating and also as a sensing element facilitating the dual functionality in the inhabited system, thereby minimize the weight and number of devices. SMA possesses an ability to remember its parent shape (austenite) and recover to it from its deformed state (martensite) at a characteristic transformation temperature. It has an unparalleled energy and power density and delivers large force and displacement. The work density of SMA is high enough to make it a very vital design factor for control applications. This chapter features the important parameters that are neccesery to be consider for designed to build an efficient control mechanism for an SMA instrumented system. This gives a broad spectrum of basic elements that are to be fixed before the developing stage, and the factors are meant to be necessarily quantified.

Keywords: Sensors, actuators, self-sensing actuation, shared sensing and actuation, variable impedance actuator

2.1 State of the Art in Shape Memory Alloy— An Introduction

Current technology has certain common desires for actuators designed for automotive, aeronautics, biomedical, robotics, and spatial applications,

---

Corresponding author: djsruth@gmail.com

Inamuddin, Rajender Boddula and Abdullah M. Asiri (eds.) Actuators: Fundamentals, Principles, Materials and Applications, (17–32) © 2020 Scrivener Publishing LLC

which is in requisite of compact and lightweight with integration compatibility, high efficient operation of force and power, and low cost. As today's technology aims towards designing of micro scales for mechanical as well as electromechanical devices. On miniaturization, classical actuators like electrical, pneumatic, and hydraulic affect a large debility power loss. These limitations led to the development of intelligent materials like piezoelectric, magnetostrictive, and shape memory alloy (SMA). When compares to these materials, SMA posseses the benefits of high strain and the highest power density enhanced with a significant amount of actuation with an extremely small envelope volume. Other dynamic inherent characteristics of SMA devices are their compliance in harsh environments, the simplicity of their actuation mechanisms, their silent and smooth motion, and their bifunctionality of sensing and actuation. These features make SMA a capable candidate among the actuators dominate in the field of aerospace and robotics.

However, it took about remarkable years for SMA from the discovery in the 1960s period to get prominent and well-known functional material, the current development could only be recognized by functional applications of SMA into medical technology (orthodontic wires, guide wires, stents) which is now commercialized widely.

2.1.1   SMA Actuators in a Feedback Control System

An SMA element operates counter to a biasing element to facilitate fixed or variable force to produce work. On actuation, the one-way SMA combined with biasing element that generates work done can be harnessed for actuator applications. SMA actuators can be used in different types of biasing configurations using either passive elements including dead mass, helical springs, cantilever strips, torsion springs, etc., or by an active element using SMA wire. By Kohl [1], it has been stated that SMAs display maximum work densities at 10^7 Jm^{-3}, i.e., 25 times more than that of the work density of electric motors when compared to conventional motors scaled to micro-size. Chaudhuri and Fredericksen [5] in the domain of robotics built a hand-actuated by SMA wires as it exhibits forces more than that of human muscle. A sigma-shaped pulley system to enhance the performance of an actuator is proposed by Hirose et al. (1986). Kuribayashi [2] proposed a control technique for the hysteresis by limiting the range of displacements and thus abridged the problem to a linear one. Hysteresis can also be reduced by employing an antagonistic SMA configuration. Further, it can be reduced by engaging direct sensors like temperature in feedback control loop. Ikuta et al. [3], proposed on estimating electrical resistance as a sensing signal for position control and experimental results revealed that the stiffness of an SMA spring is a linear function to normalized electrical resistance which led to experimenting with direct stiffness control.

2.1.2   Factors to be Considered

Assembling an SMA-based system involves the identification of three factors at the initial stage.

- Functionality in which the SMA element is employed.
- Classifying the behavioural region to operate.
- Choosing the appropriate biasing element for repetitive and cyclic operation.

2.1.2.1   Different Types of Functionality—Actuator

Having attracted property like shape memory effect and superelasticity/ pseudoelasticity of SMAs got its attention in various fields of industry. There are also few insights on interesting factors that attract SMA to be a potential candidate for a wide range of applications.

SMA instrumented systems	Types of functionality	Self-sensing actuator	
		Sensor	
		Actuator	Pseudoplastic (thermal effect)
			Pseudoelastic (mechanical effect)
	Biasing element	Unidirectional control	SMA+ spring/dead mass/flexural chassis
		Bidirectional control	OWSMA+OWSMA
			TWSMA+ spring/dead mass/flexural chassis

With high intrinsic strength of the SMA actuators, there is an advantage of being able to implement direct drive devices with smooth and noiseless operation. SMA actuated systems are made "intelligent" and their shape or stiffness is dynamically modified by embedded/bonded SMA; this helps shock absorption, which had been the primary weakness of traditional actuators. For micromanipulation control, many sensors were employed in order to achieve accuracy with the fast response time. Unfortunately, the usage of many sensors for the micro and nanodevices/systems is restricted due to their sizes, performances, and restricted degrees of freedom.

Hence, an alternative method is to use the actuators in sensing mode by capturing the variation in electrical resistance on its actuation by the concept of self-sensing method.

There are several advantages in using the self-sensing actuation (SSA) technique is that it allows a reliable reduction in expenditures and is made to be compact by eliminating expensive sensors. As the actuator is an integral part of the system, the resolution is relatively good equivalent to that of external sensors sensing signal.

This idea of recovering plastic deformation to dissipate energy in SMA by hysteretic cycles finds a broad scope in structural health monitoring systems. The presence of R-phase (Twinned martensite) is the main cause of the hysteric feature. The pseudoelasticity and damping ability of SMA is developed as a supplemental restoring element and energy dissipation actuators. The vital capability lies in the transformation from the parent phase, microstructurally symmetric cubic austenite, to the less symmetric martensitic product phase and revert back either by removing heat or by removing the load in case of pseudoelastic behavior or the evolution of R-phase in phase transformation, respectively. The austenite finish temperature is set based on the fabrication process and also the alloy composition of SMA will lead to the behavior of pseudoelastic.

Pseudoplasticity	Pseudoelasticity
Exhibits Shape memory effect	Exhibits Superelastic effect
Thermal induced phase transformation	Stress-induced phase transformation
This transformation involves M_s, M_f, A_s and A_f	This transformation occurs above A_f

In pseudoelastic, on removal of applied stress, the martensite cannot stay stable in that phase and thus reverts back to austenite. So, at low-temperature, austenite state is stable in pseudoelastic wires on contrary to the SME wires. This property is employed to develop dampers, absorbers, fasteners, clampers, and so on.

2.1.2.1.1 Variable Impedance Actuator

Actuators are a prime element for generating force and motion is a vital element of any mechatronics system. The variable compliance or variable stiffness are required in many robotic fields where human and machine interacts, unlike in classical stiff electrical drives used in industrial robotics, which demands accuracy and precision for tracking. Variable impedance actuator (VIA), depending on the external forces and the mechanical properties of the actuator, varies the stiffness from its set equilibrium position, in an unknown and dynamic environment for human-machine interaction. The adaptability of compliance in the interface with the human operator is important in applications which demand continuous force exchange. This bilateral force control needs to deliver the force continuous with accuracy for any force feedback devices for the human to exactly project the scenario on the robot slave site.

2.1.2.2 *Self-Sensing Actuation*

The traditional sensors are not ideally suitable to the micro and nano world because of their sizes, performances, and limited measurement of degrees of freedom. Henceforth, an alternative method is to use the actuators in sensing mode by capturing the variation in electrical resistance on its actuation by the concept of self-sensing method. Self-sensing is the inherent ability of a material/element to sense its own displacement/strain. SSA systems are compact as they configure a single element to sense any voltage signal preferably electrical resistance while in its actuated state.

Figure 2.1 demonstrates the SSA approach in a general control application. Apart from piezoelectric or SMA, SSA is possible with materials that intrinsically contain information about varying mechanical quantities on its actuation such as force and displacement in addition to electrical quantities like charge and electric field. In self-sensed–based control, the mechanical quantities are reconstructed by acquiring electrical quantities and the model of the transducer. So, in a generic closed-loop system, the final control element (actuator) and the transducer for feedback control

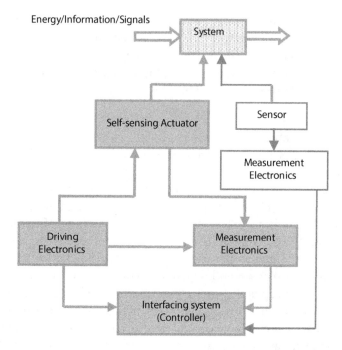

Figure 2.1 Schematics of external sensed and self-sensed actuators control.

(sensor) are the same elements performing bi-functionality. Therefore, self-sensing control application works without the necessity of any external sensor, as it is determined based on the knowledge of the model of the self-sensing actuator.

SMA actuators are a potentially viable choice for actuators that are used in a various broad spectrum of applications. SMA actuators are advantageous over piezoelectric for some applications for being able to generate larger forces, though at low frequencies and they can be made up into different shapes, like wires, tubes, sheets, and thin films Sun *et al.* [18]. SMA wires are normally used, can be embedded into the face sheet of a structure of interest, such as a helicopter blade, and can actively alter the shape of the structure in the controlled fashion (Song *et al.* [19]).

SMA can be actuated by either joules heating effect or by thermoelectric heaters. NiTiNOL (named after NiTi, which was discovered at the Naval Ordnance Laboratory). When the temperature increases in the SMA wire, the inherent properties like electrical resistance, the volume fraction of martensite and austenite, stiffness various along the phase transformation cycle. Different phase transformation mechanisms involved in heating and cooling may subsequently influence the correlation between the electrical resistance and strain output of the SMA actuator. On heating, the wire's electrical resistance first decreases due to temperature rise and vice versa. On reaching the safe heating current/transformation temperature, the wire gets contracted (increase in diameter) and the resistance decreases correspondingly, and on cooling, the SMA wire gets expanded by its biased element thereby increases the resistance. The variation of resistance in the actuation cycle can be used to specify the amount of contraction force as in Lin *et al.* [20]. The self-sensing property is the correlation between the driving voltage and the voltage drop (resistance) across the SMA wire. Electrical current is applied to the SMA wire, to contract and, hence, operated as an actuator. The change in resistance will be mapped to the consistent change in the voltage drop, as well as the structure's displacement when the wire contracts. This can be measured by using the wheat stone bridge or voltage divider method, the current is driven between the two outer points and the voltage is measured between the inner points. The voltage drop is then correlated into electrical resistance and corresponds to the SMA structure's displacements. The electrical resistance variation or the voltage drop variation signals across the SMA wire is utilized as the feedback sensing signal. This allows the determine the location of SMA-based structures in space without the need for any other external additional sensors.

2.1.2.2.1   Self-Sensing Techniques

There are three different self-sensing schemes to acquire the variation of electrical resistance in the SMA wire during actuation. Resistance feedback control is reported in kinds of literature and is focussed on SMA-spring or SMA- fixed force pairs. In unidirectional control, the operation involves opposite pulling units with sufficient stiffness element which affects the range of displacement. The range of motion can be increased using antagonistic SMA wire actuators or multi-wire actuators with self-sensing capabilities for miniaturized applications, thereby reduces the hardware complexity of placement and cost introduced by additional sensors and this is bi-directional control.

2.1.2.2.1.1   Mapping Approach

Among mapping methods, there are two approaches: polyfit approach and neural network (NN)–based approach using measurement feedback. This method involves intense open-loop experimentation.

In polyfit technique, the electrical resistance variation/resistivity/ volume fraction (an inherent property as a sensing signal) is calibrated

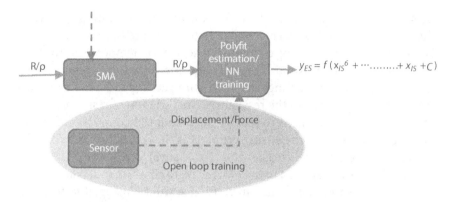

Figure 2.2 Schematics of the mapping technique.

with an external physical sensor as in Figure 2.2. The accuracy of this method is greatly reliant on the accuracy of the calibrator used and the polyfit equation is normally limited to six degrees due to software computation. Simply, it is represented as a polyfit equation of the form

$$y_{ES} = f\left(x_{IS}^6 + \cdots\ldots\ldots + x_{IS} + C\right) \tag{2.1}$$

where x_{IS} is the electrical resistance/resistivity, any internal sensor variation in the SMA wire which is mapped to get value same to that of an external sensor signal (y_{ES}). So it needs open-loop experimentation in different loading, temperature, and initial conditions to exactly get the genuine signal variation, which makes the process a tedious one.

In NN, off-line open-loop training experimentation is needed to relate the resistance-position mappings. The performance of this approach is very dependent on training the NN using several sets of input and output data obtained from the open-loop experiments. The NN controller is a hysteresis model that establishes the correlation between the applied voltage and displacement. Also, NNs, which holds nonlinear function mapping and adaptation property, are used to identify the hysteresis characteristics of mechanical systems.

2.1.2.2.1.2   MODELLING APPROACH

This modeling approach method requires sensorless feedback but liable on mathematical modeling of SMA stress-strain relation. The one-dimensional constitutive model for the SMA actuated system is developed relating the stress to the state variables like strain, temperature, and martensitic fraction. The constraints in using these models are practicality as the required model parameters must be experimentally determined. This approach has its limitations as it operable only at ambient temperature. At ambient temperature, the material properties remain mostly constant but it becomes difficult when it experiences unexpected changes.

2.1.2.3   Sensor-Actuated Isothermal (SMA)

Active sensing is a potential concept presented in this research. When a biased SMA wire is kept in an actuated state at a constant current (sensing current) and, for a varying displacement/force (measurand) applied to it, its electrical resistance changes in a linear manner; thus, the SMA is made to function as an active sensor. The value of the electrical resistance is constant for a constant current and, while in sensing mode, when an input is applied, the ER changes (increases with increase in applied load) due to the stress-induced in the wire. The selection of the sensing current depends on the stiffness of the biasing element.

2.1.3   Configurations of SMA Employed

2.1.3.1   Agonist-Antagonist Configuration

A set of actuators acts on the same rotational/linear point but opposes each other. This means that the actuators have to be synchronized to efficiently move the attached point/shaft. This arrangement has its origin in the fact that bio-inspired actuators generate force by contracting and it takes at least one other actuators to stretch them again. The number of actuators in such an agonist-antagonist set depends on the degrees of freedom. For a single degree of freedom, a pair of actuators suffices to have full actuation of the joint.

Such a configuration with SMA can be brought in a system by using it as an agonist with either a passive or active antagonist, made to function alternately in order to oppose the other. The bias is usually the antagonist to an SMA; passive antagonist can be a dead mass or passive spring and another SMA will be the active antagonist. Though SMA is non-linear, suitable design of the configuration can negate the hysteresis and exhibit linear movement. If the agonist-antagonist pair is active, then bidirectional active control can be achieved. Figure 2.3 picturizes the correlation of phase transformation in SMA with its configuration.

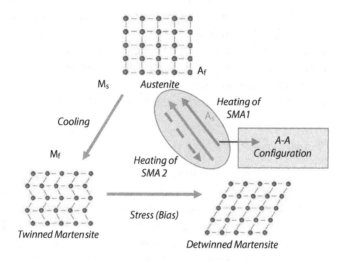

Figure 2.3 Correlation of phase transformation in SMA with its configuration.

2.1.3.2 Synergistic Configuration

In comparison to the agonist-antagonist configuration, synergistic actuators generate force in the same direction. That doesn't mean they are exactly parallel, but their actions have a similar effect on the primitive they are actuating on. If SMA wires are configured in this fashion it has an additive effect on its displacement and one such configuration is used in the design of the smart driver assistance system (SDAS). The design should be such that the displacement generated by the SMAs should be in the same direction and at the same point.

2.1.3.2.1 Configurations of Biasing Element

SMA can perform in two modes based on the shape memory effect. They are material-based one-way and two-way shape memory effect and mechanical based. The advantages of mechanical two-way actuators have a larger range of motion and than that of material two-way actuators, whereas the advantage of the material two-way actuator is simple and compact.

Figure 2.4 shows three basic types of SMA actuators using one-way SMAs.

- Figure 2.4a shows passive-biased one-way actuator. The SMA element is actuated to move in the direction of the arrow being the dead mass (W) as the biasing element.

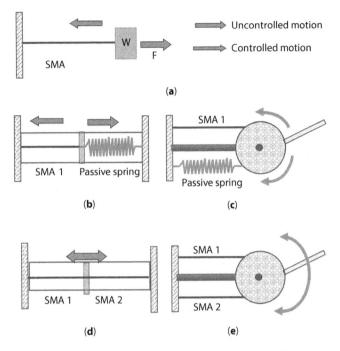

Figure 2.4 Types of SMA actuator configurations using one-way SMA wire (a) one-way actuator; (b) biased linear actuator; (c) biased rotary actuator; (d) two-way linear actuator; (e) two-way rotary actuator.

Figure 2.4b shows a spring-biased actuator, which is capable of moving the element bi-direction. The SMA element connected to the spring, and on actuation, the recovery force which is generated pull the spring, thus storing energy in it. On the removal of temperature, SMA the energy stored in the spring is released and the SMA element deforms back, thus completing the cycle. This type is configured in a rotary mode in Figure 2.4c.

On employing the passive biased actuator, only the heating of the SMA path can be controlled, whereas the cooling path is influenced by the biasing element.

• Figures 2.4d and e show a one-way shape memory effect wire that is designed to generate two-way mechanical movement with two SMA elements in antagonism. Any motion can be obtained by appropriately cooling or heating the two SMA elements. The active biased configuration involves bi-directional control.

Two-way shape memory effect-based actuator is similar to one-way actuator in shape (Figure 2.4b), while its behavior is more similar to biased actuator (Figure 2.4d).

Case Study 2.1 Governor Valve

Temperature-sensitive governor valve instrumented with NiTi thermo-variable rate spring, to control the shifting pressure in automatic transmissions by thermal actuation. This is employed in a tube with SMA spring and steel spring in a passive antagonistic configuration and which has an opening slot "S" functions to be in ON and OFF position to release pressure.

At room temperatures, the spring force of a Ni·Ti shape memory spring is lesser than that of a steel bias spring. In the martensitic state, the steel spring can compress the Ni·Ti spring, pushing the moveable piston "P" of the valve into the "closed" position. When the temperature increases, the Ni·Ti shape memory spring force is larger and it contracts, thereby making the opening slot in line with the openings and achieving the ON position. This pressure regulating valve at on and off positions are featured in Figure 2.5.

Case Study 2.2 Structural Health Monitoring

The SMA wire in pseudoelastic effect can be used as a shock-absorbing element in structural health monitoring. The SMA braces are in the frame/building structures as tendon/cables connected diagonally as in Figure 2.6. Under excitation of any vibrations, SMA braces absorb and dissipate energy through stress-induced martensite transformation (superelastic) or martensite reorientation (SME). Two designs of configurations were considered, one as in Figure 2.6a is that eight damper devices made of the

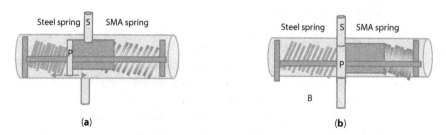

(a)   (b)

Figure 2.5 Represents valve: (a) un-actuated-martensite state "OFF"; (b) actuated-austenite state "ON".

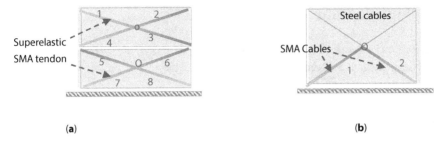

(a)   (b)

Figure 2.6 SMA braces: (a) un-actuated-martensite state; (b) actuated-austenite state.

SMA wires and steel wires are diagonal with the center "O" installed in a two-story structure. Another is to have combined steel-SMA–type braces for the frame structure with 2 of each. The dynamic characteristics and transient response hybrid tendons combinations of steel and Nitinol in the structure can be used for vibration control of coastal structures and the proper tendon geometry and there is pre-strain dependency in the dynamic response of the frame, respectively.

Case Study 2.3 Medical Staples

Shape memory orthopaedic staples are biocompatible alloys, under the shape memory effect accelerates the healing process of bone fractures. The shape memory staple, in its opened shape, OFF state in martensitic phase is driven into each side of the fractured site as represented in Figure 2.7. As the staple gets heated up to body temperature, this staple tends to contracts, compressing the separated part of bones and closes the gap between thems. This allows the stabilization with constant force accelerates healing, decreasing the time of recovery.

(a)   (b)

Figure 2.7 Medical SMA staple: (a) un-actuated-martensite state; (b) actuated-austenite state.

Acknowledgment

The contributors are hosted and supported at Robert Bosch Centre for Cyber Physical Systems, Indian Institute of Science, Bengaluru-560 012.

References

1. Kohl M., Microactuators. In: *Shape Memory Microactuators. Microtechnology and MEMS*. Springer, Berlin, Heidelberg, 2004.
2. Kirubayashi, K., Improvement of the response of an SMA actuator using a temperature sensor. *Int. J. Rob. Res.*, 10, 13–20, 1991.
3. Ikuta, K., Tsukamoto, M., Hirose, S., Shape Memory Alloy Servo Actuator System with Electric Resistance Feedback and Application For Active Endoscope, 1988.
4. Hirose, S., Ikuta, K., Umetani, Y., Development of a shape memory alloy actuator. Performance assessment and introduction of a new composing approach. *J. Adv. Robot.*, 3, 1, 3–16, 1989.
5. Chaudhuri, P. and Fredericksen, D.H., Robot hand with shape memory musculature, IBM Tech, Disclosure Bulletin, 28, 302–303, 1985.
6. Nakano, Y., Fujie, M., Hosada, Y., Hitachi's robot hand. Robotics Age, July 6, 7, 18–20, 1984.
7. Honma, D., Miwa, Y., Iguchi, N., Micro robots and micro mechanisms using shape memory alloy. In: *Third Toyota Conference on Integrated Micro Motion Systems, Micro-machining, Control and Application*, Nissin, Aichi, Japan, October 1984.
8. Bergamasco, M., Salsedo, F., Dario, P., A linear SMA motor as direct-drive robotic actuator, IEEE International Conference on Robotics and Automation, Scottsdale, Arizona, USA, 618–623, 1989.
9. Hesselbach, J. and Stork, H., Electrically controlled shape memory actuators for use in handling systems Proceedings of the 1st International Conference on Shape Memory and Superelastic Technologies, Monterey, CA, USA pp. 277–282, 1994.
10. Stoeckel, D. and Tinschert F., Temperature compensation with thermovariable rate springs in automatic transmissions. SAE technical paper series: SAE, 1991.
11. Song, G., Ma, N., Li, H.N., Applications of shape memory alloys in civil structures. *Eng. Struct.*, 28, 1266–1274, 2006.
12. Machado, L.G. and Savi, M.A., Medical applications of shape memory alloys. *Braz. J. Med. Biol. Res.*, 36, 6, 683–691, 2003.
13. Josephine Selvarani Ruth, D. and Dhanalakshmi, K., Shape Memory Alloy Wire for Force Sensing. *IEEE Sensors*, 17, 4, 967–975, 2017.

14. Josephine Selvarani Ruth, D. and Dhanalakshmi, K., Shape memory alloy wire for self-sensing servo actuation. *Mech. Syst. Signal Pr.*, 83, 36–52, 2017.

15. Josephine Selvarani Ruth, D., Dhanalakshmi, K., Sunjai Nakshatharan, S., Bidirectional angular control of an integrated sensor/actuator shape memory alloy based system. *Measurement*, 69, 210–221, 2015.

16. Josephine Selvarani Ruth, D., Dhanalakshmi, K., Sunjai Nakshatharan, S., Interrogation of Undersensing for an Underactuated Dynamical System. *IEEE Sensors.*, 15, 4, 2203–2211, 2015.

17. Josephine Selvarani Ruth, D., Chapter 9 Robotic Assemblies Based on IPMC Actuators, Springer Science and Business Media LLC Springer Nature Switzerland AG, 2019.

18. Sun, L., Huang, W.M., Ding, Z., Zhao, Y., Wang, C.C., Purnawali, H., Tang, C., Stimulus-responsive shape memory materials: A review. *Mater. Des.*, 33, 577–640, 2012.

19. Song, G., Chaudhry, V., Batur, C., Precision tracking control of shape memory alloy actuators using neural networks and a sliding mode based robust controller. *Smart Material Struct.*, 12, 223–231, 2003.

20. Lin, C. M., Fan, C.H., Lan, C.C., A shape memory alloy actuated microgripper with wide handling ranges. IEEE/ASME International Conference on Advanced Intelligent Mechatronics (AIM) 2009, 14–17 July 2009 Singapore, 2009.

Actuators in Mechatronics

Akubude, Vivian C.[1]*, Ogunlade, Clement A.[2] and Adeleke, Kehinde M.[3]

[1]Department of Agricultural and Bioresource Engineering, Federal University of Technology, Nigeria
[2]Department of Agricultural Engineering, Adeleke University, Ede, Nigeria
[3]Department of Mechanical Engineering, Adeleke University, Ede, Nigeria

Abstract

Mechatronics system or robotic is a present day technology which encompasses the incorporation of mechanical system, electronics and computer control in solving complex problem, increasing speed and precision while reducing stress for the human race. Motion and control are inevitable in mechatronics system and the essential element of the system that handles such is the actuation system. This chapter discusses the components of the actuation system, types of actuators and applications of actuators.

Keywords: Mechatronic, robotics, actuators, actuator types, actuator application

3.1 Introduction

The advancement in technology has paved way for several forms of actuators with wider applications. The need for accomplishment of task via the use of machines to reduce human fatigue, save time, and increase productivity with high precision has given room for constant improvement and advancement in technology. Mechatronics systems are product of such advancement in technology which has helped in solving a lot of present day challenges. Every mechatronic system has an actuating unit which is responsible for motion and controls.

*Corresponding author: akubudevivianc@gmail.com

Inamuddin, Rajender Boddula and Abdullah M. Asiri (eds.) Actuators: Fundamentals, Principles, Materials and Applications, (33–44) © 2020 Scrivener Publishing LLC

3.2   Mechatronics System

Mechatronics is a multidisciplinary field that involves the synergetic integration of mechanical engineering (mechanical elements, machines, precision mechanics) with electronics (microelectronics, power electronics, sensor, and actuator technology) and intelligent computer control (systems theory, control and automation, software engineering, artificial intelligence) in the design and manufacturing of products and processes [1]. According to Robert and Ramasubramanian [2], mechatronics is the synergetic integration of mechanical engineering with electronics and intelligent computer control in the design and manufacturing of industrial products and processes. The key elements of mechatronics are

(1)   Physical systems modeling
(2)   Sensors and actuators
(3)   Signals and systems
(4)   Computers and logic systems
(5)   Software and data acquisition

3.3   Structure of Mechatronics System

Mechatronic encompasses different fields including mechanical, systems control, electrical, and electronics engineering [3, 4] to develop unifying design and various elements as shown in Figure 3.1.

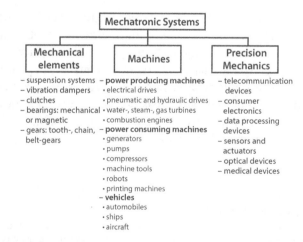

Figure 3.1 The mechatronics system [5].

3.3.1   Elements and Structures of Mechatronics Systems

The principal elements of mechatronic system including mechanical, control interface, electrical, electromechanical, and computer elements (Figure 3.2) form the base structure of mechatronics as shown in Figure 3.3. A closer look at relations between the system, the sensors, the information processing, and the actuators shows that a description of the relations

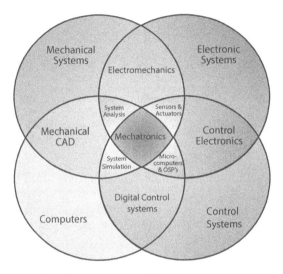

Figure 3.2 Elements of mechatronics system [6].

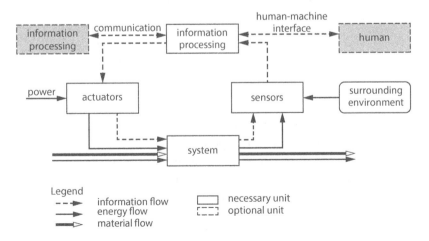

Figure 3.3 Base structure of a mechatronics system [7].

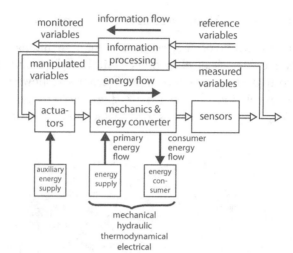

Figure 3.4 Energy and information flow within a mechatronic system [5].

using flows is useful (which include material, energy and information) as presented in Figure 3.4.

3.4   Actuators

Actuators are used to produce action or motion; they together with sensors and digital electronics are increasingly integrated into mechanical systems. They play a major role in mechatronic systems and are essential for efficient design process.

3.4.1   Components of Actuation System

The two major components of actuation system are:

i.   Power Amplification and Modulation Stage: this converts the control signal into appropriate signal that delivers the required input power to the energy conversion unit. This element consists of a power electronic circuit in electrical drives, such as, providing the appropriate high power switching to the electrical drive.

ii.   Energy Conversion Stage: this converts energy and produces work.

3.4.2   Types of Actuators

The common types of actuators in automation include: Pneumatic, Hydraulic, Electromechanical, and Mechanical Actuators.

  i.   Pneumatic Actuators: this is commonly used in manufacturing industries. It uses compressed air to transmit power between various elements. It is used for energy transfer devices, particularly in the case of linear motion actuators. It is characterized by low weight, compact size, low cost availability, low maintenance, ability to be used under extreme temperature and low power capability [6]. Pneumatic actuators exhibit similarity with hydraulic actuators in terms of function and construction. They comprises of Compressors (acts as power sources), Valve (controls the direction and regulates the air flowing to actuating system), Pipes (connects various elements of actuating system), Air treatment unit (regulates flow of compressed air), Sensors, and Controller (monitors whole system to provide required output) as presented in Figure 3.5.

  ii.   Hydraulic Actuators: It uses fluid power to produce mechanical force. The mechanical motion gives an output in terms of linear, rotary, or oscillatory motion. Hydraulic actuating system consists of sensors, controller, and

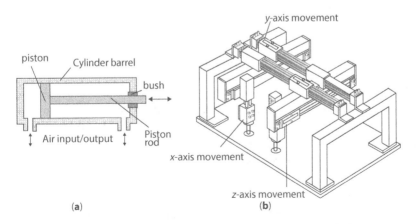

Figure 3.5 Pneumatic actuation: (a) Pneumatic Cylinder; (b) A pick-and-place pneumatic system [6].

actuating components [8]. This system can handle heavy load with less input. The main parts of the hydraulic actuator include: Control device (controls whole actuating system by providing signals), Valve (regulates direction, fluid flow rate, and pressure), Pump (pressurizes fluid to required level), Sensors (supplies feedback signal), reservoir, Pipes, filter and accumulator.

iii. Electromechanical Actuators: they change electrical into mechanical energy. It possesses high response speed, can be maintained and controlled easily, low cost, and a clean energy source though it is susceptible to mechanical failure due to moving parts. A very good example of this type of actuators is electric motor (which can be DC motors, AC motors, and stepper motors).

iv. Mechanical Actuators: this kind of actuator aids the transformation of motion into other forms using pulleys, belts, gears, chains, or rigid links. Belts and chains transmit torque from one pulley to another pulley at a specific distance; gears are used to transform rotary motion to another where torque variation is required. They are difficult to implement in complex motion condition.

3.5 Actuator Components

Actuator components can be grouped into three headings, namely, hydraulics, pneumatics, and electric motors.

3.5.1 Hydraulics and Pneumatics

The two have similar components only that hydraulic actuators use liquid as driving force while pneumatic actuators make use of air as driving force. They are otherwise known as mechanical actuators. The most important components include regulators, actuator cylinder, speed valves, solenoid valve, reservoir, and hand valve. They are discussed below.

• Regulators
Regulators are devices used to control circuit pressure. The pressure being a measure of force action on a given unit area in the circuit makes the device to be able to control the actuating force in the system. They are

connected with components that respond to change in the downstream fluid pressure mechanically with an indicator gauge. As long as the pressure supplied by the reservoir is greater than required circuit pressure, the regulator attempts to maintain automatically, a preset pressure within the pneumatic circuit system.

- Actuator Cylinder

These are devices that are used in applying either pulling or pushing forces. They are designed to perform either linear or rotary reciprocating motions. The two forms of the actuators are majorly found in thousands of configurations depending on the design. The simplest configuration comprises the pistons of various lengths of stroke and bore depending on the nature of energy to be dissipated in the required system. They are specified mostly either as being powered in one direction (single acting) or two opposite directions (double acting). It is generally observed that a single acting spring return cylinder consumes less fluid and hence more economical [9].

- Speed Valves

Speed valves as the name implies control the speed of fluid flowing in or out of a pneumatic circuit. They are otherwise known as flow valves. The flow being a measure of fluid volume passing through the circuit at a given period of time makes the valves to be able to control the volume of fluid at any given instance. The device comprises a screw needle mounted across a cylindrical pipe with inlet and outlet openings. The screw needle, when adjusted inwards decreases the flow rate and vice-versa. The higher the flow rate, the faster the pneumatic actuator operates.

- Solenoid Valve

These are electrically operated devices designed to control both the flow and the direction of the pressurized fluid in and out of pneumatic actuators. They designed to work with a preferred control system using Programmable Logical Controller (PLC), Programmable Interface Derivative (PID), or predetermined constant output controller such as NE 555 timer Integrated Circuit (IC) to give the desired output. Solenoid valves are designed to operate as either mono stable or bi-stable. The mono stable connection spring return to a default condition either on or off while the bi-stable connection has no default or preferred condition and thus remains where it was last positioned either on or off.

Generally, pneumatic valves can be operated by hand (mechanical), fluid (piloted), or solenoid (electrical) operated.

- Reservoir

The air reservoir operates with the principle of air compressors. It stores pressurized air to actuate components of pneumatic circuit. It is used to supply air to the system at any instance. A typical air reservoir consists of

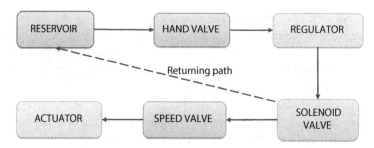

Figure 3.6 A typical arrangement of a pneumatic or hydraulic circuit actuator components.

an electrically operated air pump, a shell, control valve, pressure sensor, and gauge.

While the air reservoir is designed to either trap the air from the atmosphere or recycles the used air in the system, hydraulic liquid is designed only to be recycled into the reservoir and reused to avoid wastage. To achieve this, an additional non-return valve is added to the exhaust port of the solenoid valve.

• Hand Valve

This is an essential component that is operated manually, used to shut the flow of fluid into the entire circuit for safety. When the fluid pressure is turned off, the circuit pressure is automatically vented and renders the fluid circuit safe.

The schematic diagram in Figure 3.6 shows a typical arrangement of both the hydraulic and pneumatic circuit system.

3.5.2 Electric Actuators

The basic components of electrical actuators include electric motors and relays.

• Electric Motors

Electric motors are actuators that convert the energy received from electricity to mechanical torque. The common components include the rotor (armature), copper windings, stator magnet, commutator, carbon brush, motor shaft, and motor housing. They can be grouped into three types based on their usage and the design, namely, DC servomotors, AC motors and Stepper motors and are discussed as follows.

(i)   DC Servomotors: These are special types of DC motors that use a feedback loop to control its speed. They are conveniently used by direct current and provide a linear torque-speed relationship. They are widely used in mechatronics because of ease of control from programmable controllers.

(ii) AC Motors: These are motors that run on alternating current and are widely used in the industry. They possess high power supply and easier to maintain.

(iii) Stepper Motors: These are motors that provide an exact movement form of discrete angular displacement (step angles). Each angular step is actuated by a discrete electrical pulse. They are widely used in open-loop control systems.

A typical example of electric motor is shown in Figure 3.7.

• Relays

These are electrically operated switches used in the control circuit together with other components such as transistors, diodes, etc., with the current received from microcontroller system. It is employed to switch current on in one electrical circuit or off in another circuit. When there is a flow of current into the circuit, the coil is energized to switch to an alternative

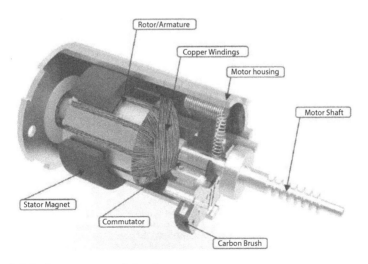

Figure 3.7 Basic components of electric motor.

terminal and when de-energized, it switches back to its initial terminal. They can perform four different delay timing tasks, namely:

(i)   Normally open, Time closed,
(ii) Normally closed, Time open,
(iii) Normally open, Time open, and
(iv) Normally closed, Time closed.

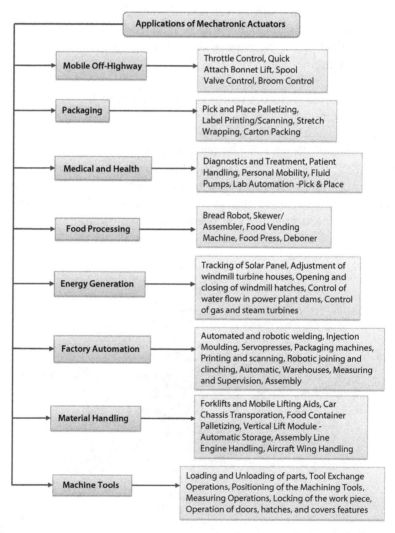

Figure 3.8 Applications of mechatronics actuators.

3.6 Applications of Actuators in Mechatronics System

There are several applications of actuators in mechatronic systems; most of which are used to provide desired movement in different dimensions. The common ones are highlighted below.

- Robotics: A common example of actuators is in robots where servomotors are used to drive links (joint) of robot to provide a desired motion as controlled by softwares.
- Material Handling Equipment: Actuators are used in various forms to drive conveyors and other material handling equipment to the specified locations.
- Processing Plants: Complex plants and machineries that are widely used in the industries found indispensable, the actuators to carryout special tasks. Several types of motions are carryout in different forms to move substances in different forms, packaging, and inventory.

The schematic diagram shown in Figure 3.8 according to Intelligent Motion Control Limited [10] gives the summary of other common applications of actuators and where applicable.

References

1. Isermann, R., *Mechatronics design approach in the mecharonics handbook*, R.H. Bishop (Ed.), pp. 27–42, CRC Press, London, 2002.
2. Bishop, R.H. and Ramasubramanian, M.K., *What is mecharonics in the mecharonics handbook*, R.H. Bishop (Ed.), pp. 27–42, CRC Press, London, 2002.
3. Bradley, D.A., Loader, A., Burd, N.C., Dawson, D., *Mechatronics, Electronics in products and processes*, Chapman and Hall Verlag, London, 1991.
4. Karnopp, D.C., Margolis, D.L., Rosenberg, R.C., *System Dynamics: Modeling and Simulation of Mechatronic Systems*, 4th Edition, Wiley, Hoboken, New Jersey, United States, 2006.
5. Isermann, R., Mechatronic systems: Concepts and applications. *Trans. Inst. Meas. Control*, 22, 1, 2000.
6. MFE 3004, Actuation in Mechatronic Systems. Retrieved online from https://www.meprogram.com.au/wpcontent/uploads/2018/03/MFE3004 AdditionalPT5.pdf, 2019.

7. Březina, T. and Singule, V., Design of Mechatronic Systems. BRNO University of technology faculty of Mechanical Engineering Institute of Solid Mechanics, Mechatronics and Biomechanics, 2006.

8. WPC, Hydraulic Actuators. http://www.wpc.com.au/actuation-controls/hydraulic-actuators, 2019.

9. Ponomareva, E., Hydraulic and Pneumatic Actuators and their Application Areas". http://www.inmoco.co.uk/actuator_applications, 2006.

10. Intelligent Motion Control Limited. Actuators Applications. Web link: http://www.inmoco.co.uk/actuator_applications, 2011.

4

Actuators Based on Hydrogels

Alexis Wolfel[1,2] and Marcelo R. Romero[1,3]*

¹Universidad Nacional de Córdoba, Facultad de Ciencias Químicas, Departamento de Química Orgánica, Córdoba, Argentina
²Consejo Nacional de Investigaciones Científicas y Técnicas (CONICET), Instituto de Física Enrique Gaviola (IFEG), Córdoba, Argentina
³Consejo Nacional de Investigaciones Científicas y Técnicas (CONICET), Instituto de Investigación y Desarrollo en Ingeniería de Procesos y Química Aplicada (IPQA), Córdoba, Argentina

Abstract

Many electronic actuators of different dimensions, such as motors, relays, pumps, and mixers, have evolved rapidly, and many are commercially available. Its application into soft environments such as biological tissues is desirable to develop automated biomedical devices. However, they present innate difficulties. Limitations are mainly due to its structural rigidity and chemical composition. To address this need, soft-actuators are under development. These devices have great mechanical and chemical similarity with biological tissues and, as a consequence, many have a high biocompatibility. Soft-actuators are capable of mimicking the behavior of muscles or other tissues that are involved in locomotion in biological organisms. Most of them are hydrogels which are polymers structured in three-dimensional networks, capable of harboring large amounts of water. Hydrogels can reversibly change its volume up to 50, 100 times, or more, from a dried or collapsed state to a hydrated state. This shape shifting can be triggered by external stimuli such as: pH, temperature, humidity, light, presence of specific molecules, among others. In this work, we present important concepts of the properties, synthesis and characterization of hydrogels. Subsequently, the main characteristics of actuators are detailed, and the results of these works are explained with to the previously described properties.

Keywords: Hydrogels, soft-actuators, tissue-mimicking, biomaterials, smart-materials, artificial-muscles, biodevices

*Corresponding author: marceloricardoromero@gmail.com

Inamuddin, Rajender Boddula and Abdullah M. Asiri (eds.) Actuators: Fundamentals, Principles, Materials and Applications, (45–74) © 2020 Scrivener Publishing LLC

4.1   Introduction

Thanks to the rapid advance of microelectronics, particularly of microprocessors and its ability to handle and process data, robotics has become the next technological challenge. In this sense, the development of versatile actuator systems, small and compatible with different environments and with precise movement control, is essential requirements to optimize automated systems of high complexity. The integrations of automated systems with soft environments such as biological tissues promoted the obtainment of soft-actuators, which are capable of mimicking the behavior of muscles or other tissues that are involved in locomotion in biological organisms (Figure 4.1).

Hydrogels (HGs) are crosslinked polymer networks that have the ability to absorb and retain large amounts of water or biological fluids and possess numerous characteristics in common with biological tissues [1]. These materials have pores that allow to enter large amounts of water. However, polymer chains are linked together, preventing the complete dissolution of the material. These bonds can be physical or covalent and the origin of the polymer can be synthetic, natural, or a combination of the above (semi-synthetic). The high water absorption capacity is achieved by the incorporation of hydrophilic functional groups (GF) in the network (for example, $-OH$, $-NH_2$, $-CONH$, $-COOH$, $-SO_3H$, among others) [2].

Some hydrogels, considered superabsorbent, can absorb up to a thousand times their dry weight, by mass of water. This large capacity has been exploited in the development of various applications, for example, for use in disposable diapers or in the improvement of water retention in soils [3]. The polymeric nature of the gels gives them chemical versatility, allowing a wide variety of functional groups to be introduced. In addition, the

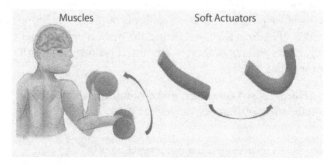

Figure 4.1 Soft-actuators are capable of mimicking the behavior of muscles or other tissues.

aqueous environment inside involves a liquid/solid interface that is an opportunity for the development of applications such as catalysis supports, devices for environmental remediation [2], sensors and actuators [4, 5] highly water swellable polymer networks capable of converting chemical energy into mechanical energy and vice versa. They can be tailored regarding their chemical nature and physical structure, sensitiveness to external stimuli and biocompatibility; they can be formed in various structures and integrated into (micro-, among others. One of the main characteristics of hydrogels is their similarity with biological tissues. This has encouraged the development of a significant number of applications oriented to the area of biomaterials. Among them, hydrogels for the controlled release of bioactive agents, contact lenses, patches for wound healing, platforms for tissue or organ repair (tissue engineering), separation of biomolecules or cells, among others [6].

Intelligent materials are those capable of sensing stimuli in their environment and respond in a functional and predictable manner, by changing their physical-chemical properties [7]. Smart hydrogels respond to changes in their environment to modify their properties in a functional way. In general, they are able to change their degree of swelling in response to the external stimulus, moving from a collapsed (or less swollen) state to a state of greater swelling (Figure 4.2). Due to this characteristic, the hydrogels are linked to the presence of a solvent for performing a mechanical work, and it is an advantage when the composition of the liquid medium is similar to the interstitial liquid present in biological tissues. Hydrogels can respond to a wide variety of stimuli: changes in temperature or pH, electrical stimuli, salinity, presence of enzymes, redox potentials, light, mechanical stress, magnetic fields, presence of an analyte, among others. In addition,

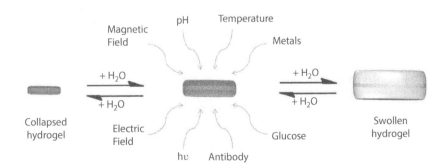

Figure 4.2 Smart hydrogels: stimuli-responsive swelling behavior. Several stimuli can trigger a change in smart hydrogels' properties, thus providing a sensor useful for actuators' design.

to their swelling properties, they can have another wide variety of intelligent responses, such as: self-curing capacity, shape memory, controlled release of bioactive drugs, and degradation on demand in biological environments. Many of the applications under development, such as operation in complex environments and complex responses, are based on hydrogels that combine responses to multiple stimuli [6, 8, 9].

4.2   Hydrogel Synthesis

A characteristic that must be known in detail for the development of hydrogel-based actuators is the synthesis of these polymers. This fundamental characteristic is often overlooked in the articles of actuators and it is tedious for a researcher who is not in the area of organic synthesis to review articles of this type to find out how to carry out or modify the synthesis of a hydrogel. For this reason, this section details the most relevant methodologies for the synthesis of hydrogels with the introduction of some practical examples to carry out it without difficulties.

Hydrogels, strongly water absorbent materials, are obtained from mainly (but not only) hydrophilic precursors. Occasionally, the combination with hydrophobic precursors is used to improve the mechanical properties of the gel and control its degree of swelling [10, 11]. The precursors can be of natural or synthetic origin. Natural polymers are frequently used to take advantage of their intrinsic properties (some from renewable, biodegradable, biocompatible, economic sources, etc.), but introducing synthetic modifications that improve the mechanical properties or functionality of the material.

Hydrogels can be synthesized in various ways, generally centered on the union or crosslinking of pre-existing polymeric or oligomeric chains or by polymerization and crosslinking of monomer units, to give a three-dimensional network. Depending on the nature of the cross-linking, hydrogels can be classified into [2]:

4.2.1   Chemically Crosslinked

They have covalent crosslinks between polymer chains. These structures can be achieved by polymerization of monomers (generally mono- or di-functional) and cross-linking agents (di- or poly-functional) of low molecular weight, by radical polymerization or by complementary GF condensation. They can also be obtained by cross-linking of pre-existing polymer chains that possess complementary GF for a reaction (polymer/

polymer cross-linking, e.g., cross-linking of a polymer with GF aldehyde with another with GF amino) or by reacting the side GFs of the polymer with molecules of low molecular weight with complementary GF (cross-linking agents). Some polymers can be covalently crosslinked by the incidence of high energy radiation.

4.2.2   Physically Crosslinked

They have physical or secondary bonds. The polymers can bind through ionic interactions, hydrogen bridge interactions, complex formation, hydrophobic interactions, amorphous or crystalline domains, supramolecular inclusion complexes, among others.

Both the chemical nature of the precursors used, such as junctions or cross-links, will define the swelling properties, mechanical properties, and stimulus response of the hydrogels obtained. Although the use of precursors of natural origin is preferable due to its lower environmental impact, hydrogels obtained by polymerization of different synthetic monomers generate more versatile materials, allowing to design and control their properties by controlling the composition and functionality of the monomers. Hence, among the many hydrogel synthesis techniques, obtaining by free radical polymerization is one of the most used strategies. This technique focuses on the radical addition reaction of mono-vinyl monomers, crosslinked with di- or poly-vinyl agents (Figure 4.3). The reaction is

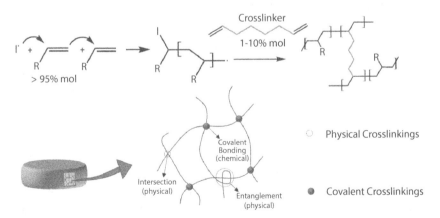

Figure 4.3 Hydrogel synthesis by free radical polymerization of vinyl monomers. A radical initiatior triggers a chain reaction for the polymerization of monomers and crosslinkers yielding a tridimensional network.

promoted by the use of a radical initiator, which generates the first radicals in the reaction medium. These will be added to the double bonds present in the vinyl or crosslinking monomers, generating new radicals of increasing molecular weight and producing a chain reaction that leads to the formation of high molecular weight polymers, covalently crosslinked. To start the polymerization, various initiators of the radical reaction are used. Some of them form radicals from their thermal decomposition (for example, peroxides or azo compounds), others from irradiation with light of a certain wavelength (photo initiators), and some through an oxide-reduction reaction [2]. The most widely used are persulfate salts (ammonium, sodium, or potassium persulfate) which decompose generating radicals, when heated in aqueous solution (usually heated above 60°C). Occasionally, persulfate salts are used together with alkyl-substituted amines that allow accelerated decomposition of persulfate, promoting rapid generation of radicals through an oxide-reduction mechanism. Redox initiation occurs more quickly because it has a lower reaction energy barrier, so it also allows polymerization to be carried out at temperatures lower than those used in thermal decomposition of persulfate. The most commonly used amine, for having demonstrated greater effectiveness, is tetramethylethylenediamine (TEMED) [12]. Regarding the reaction medium, since hydrophilic precursors are mainly used in the synthesis of hydrogels, in general, the synthesis is carried out by polymerization in aqueous solution.

4.3   Experimental: Radical Polymerizations

4.3.1   General Recommendations

Radical reactions are inhibited by the presence of oxygen. Therefore, it is recommended to perform a nitrogen bubble for a few minutes to displace the air present in the reaction solution. In general, it is convenient to avoid the contact of air in the anoxic solution, protecting it with a cap. Then, the solution should be quickly located in a mold with cap to obtain the desired morphology. On the other hand, the quality of water used is important, so at least distilled water should be used, since the presence of impurities in the solvent significantly affects the performance of the reaction. In general, to carry out this reaction, it is placed in a container: pure water, monomers, and initiator. The monomers (MM) that have vinyl groups in their structure are weighed to an analytical concentration of 1–1.5 M. Then, to initiate the radical reaction, one of the methods described below can be used. The amount of initiator to use usually varies between 0.5% and 5% (in moles with respect to MM):

- # 1- Thermal radical initiation: the most commonly used radical initiators in aqueous reactions are persulfates. For example, ammonium persulfate (APS), potassium persulfate (KPS), or sodium persulfate (NaPS) can be used. These substances are hygroscopic, so they must be protected from moisture during storage. After incorporating the initiator into the reaction mixture, the container or mold containing the solution should be placed in a heating bath between 50°C and 70°C. Depending on the composition, the reaction may form macroscopic gel from a few minutes to a few hours. In general, it is recommended that during the reaction, the curing is carried out to protect the ambient air.

- # 2- Radical initiation at room temperature: for start the reaction at room temperature, an accelerator is added. In general, tetramethylenediamine (TEMED) is used, which is added in the same concentration as the persulfate initiator. In this case, after the addition of both components, the reaction will begin quickly (less than one minute). Given the high speed of initiation, many times the time to load a mold or form a film may be insufficient. For this reason, a 5% solution of TEMED in water is usually used, which must be added once the other components are already soluble and the solution is free of dissolved air. The TEMED solution can be directly injected when the reaction mixture is already located in the required mold or an immediate transfer to the mold must be made after the accelerator has been added. It is recommended that the reaction be carried out to protect the ambient air.

- # 3- Radical initiation with UV light: polymerization is promoted from a molecule capable of forming radicals when irradiated with ultraviolet light. The light commonly used is that provided by a mercury lamp, whose main emission is around 254 nm. The lamp is placed inside a dark chamber, and protection on the body and mainly the eyes should be used when performing this method. The advantage of this initiation is the short time required to achieve the reaction, which is almost in a few seconds. The disadvantage is that the penetration of UV light into the materials is not very deep, so it is not recommended if gels more than 1–2 mm thick are desired. Some of the initiators used are: 2,2 dimethoxy-2-phenylacetophenone (DMPA) or phenylbis(2,4,6 trimethylbenzoyl) phosphine oxide (PBPO), frequently added in a concentration of 0.1%–1% in mol with respect to the concentration of monomers. In this reaction, the lamps, after turn on, have an intensity variation that stabilizes in about 10–20 min. As a consequence, it is advisable to stabilize the lamps before exposure of the sample to light, to achieve greater reproducibility of the results. This type of reaction allows the use of masks to achieve precise control of the structure of the material. On the other hand, the material can be deposited using a 3D printer, to achieve very complex structures.

- # 4- Radical initiation with visible light: It is similar to UV radical reactions. It has the advantage that the visible light used can be supplied by LED light source. Thus, the use of eye protection for radiation is harmless and unnecessary. However, it has the disadvantage that ambient light could initiate the reaction, so the reaction system must be protected from external spurious light. Typical initiators used for this reaction are ethyl α-bromophenylacetate (EBPA), 2-bromopropane nitrile (BPN), or phenylbis(2,4,6 trimethylbenzoyl) phosphine oxide (PBPO) (IRGACURE 819) at higher concentration than UV. Other used initiators are camphorquinone together with N, N dimethyl-p-toluidine, 2-ethyl-methyl benzoate, or N-phenylglycine.

When laser is used as light source, methylene blue irradiated at 800 nm can initiate the reaction. This method has the advantage of the excellent dimensional controls being able to collimate the laser light to a few micrometers.

4.4 Mechanical Properties

In hydrogels, the mechanical properties can be modified by changing the type and degree of crosslinking of the hydrogel. For example, an increase in gel stiffness can be achieved by increasing its degree of crosslinking. As a consequence, this could generate a limitation in the swelling capacity of the material and make it more brittle [1]. It is also possible to modify the mechanical properties by altering the nature of the interactions between the polymer chains, for example, generating domains of hydrophobic associations. The optimization of the mechanical properties must be carried out according to the actuator development and optimization (as required by a more elastic, viscous, rigid material, etc.) and considering that these are closely linked to the swelling properties.

4.4.1 Characterization by Rheology

Ideally, against an external deformation force, a material could behave as an ideal solid or an ideal liquid. In an ideal solid, the applied stress force would result in an elastic and instantaneous deformation of the material, which completely return after the removal of the stress. This would correspond to an ideal elastic behavior, equivalent to the ideal behavior of a spring, mathematically represented through Hooke's law. If the material behaves like an ideal Newtonian liquid, all the applied force would generate an inelastic and irreversible deformation of the material, where the energy used would be lost by an inelastic deformation. Actually, hydrogels have an intermediate behavior between these two ideal materials, so it is named a viscoelastic behavior. To be precise, part of the energy applied to the material is stored elastically and another part is viscously dissipated. But this depends not only on the nature of the material, nor on the magnitude of the force, but also on the speed with which that force is applied.

Materials could behave as ideal solids or liquids depending on the time scale of application (speed or frequency) of mechanical stress (external force). A hydrogel could act elastically (like a rubber) before a deformation force, responding instantaneously to the stress applied and recover completely after the stress is removed; or it could deform inelastically, like a viscous liquid if the same force were applied at a slower speed or for a longer period of time. This period of time would allow the relaxation of the material, with respect to the tension applied. Therefore, it is said that the viscoelastic behavior is time-dependent and a certain elastic behavior will be observed combined with a viscous behavior, whose relationship depends on the speed at which the tension was applied [13]. The study of the viscoelastic behavior of a hydrogel through the application of external forces at different times (frequencies) allows to obtain information about the interactions (covalent or physical) in the material, the mobility of the chains, the degree of crosslinking, among others. The oscillatory rheology consists in measuring the response of the material to mechanical stress when a shear deformation of sinusoidal behavior (with respect to the deformation angle) is applied. The hydrogel is placed between two parallel plates that apply the deformation and measure the response of the material.

The viscoelastic behavior at a certain frequency (ω) is characterized by measuring the complex modulus (G *) of the material. This will have an elastic contribution, reflected in the storage module (G', also called elastic module) and viscosity, reflected in the loss module (G', also known as viscous module) as shown in equation (4.1):

$$G^*(\omega) = G'(\omega) + i. G''(\omega) \qquad (4.1)$$

The main characterization tests of hydrogels consist in determining their behavior before a sweep of the amplitude of the deformation (1) and a frequency sweep (2) [13].

In the first, the percentage of deformation imposed on the hydrogel (rotational displacement of the material) is varied, at a constant frequency (Figure 4.4). Through this test, it is possible to determine the Viscoelastic Linear Range (VLR) of the hydrogel. In the VLR, the values of G' and G" are independent (approximately constant) of the amplitude of the deformation imposed. After exceeding the VLR (exceeding the "critical point"), the elastic response of the material against deformation decreases, indicating an increase in energy dissipation through inelastic deformation (decreasing G' and G").

This test allows to observe differences in the extent of the VLR between different hydrogels to know how much deformation the material can withstand responding elastically and what is the moment at which inelastic deformations begin to occur due to displacements between the chains or internal fractures of the material. The extension of the VLR at a given frequency is linked to the type of interactions and degree of crosslinking of the material. A more elastic material will have an extensive VLR, while a more brittle will be in elastically deformed, by fractures, after minor deformations. However, as mentioned earlier, the viscoelastic behavior of hydrogels is time-dependent. That is to say, given the same strain amplitude (same mechanical stress energy), the hydrogel can behave more elastically or more viscously, depending on the time and speed of force application. If the force is applied for a long time, it may be inelastically deformed by phenomena of relaxation of polymer interactions. This allows valuable information to be obtained from the strain frequency sweep test. In this test, the hydrogel is deformed at a constant amplitude, but the frequency (speed) of deformation application is varied (Figure 4.5). This study makes it possible to compare the G' and G" of hydrogels, with different degrees

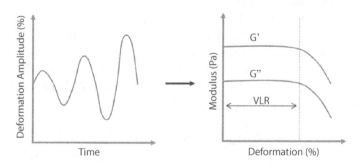

Figure 4.4 Rheological test: Amplitude sweep of deformation. The determination is performed under a constant deformation frequency. It is useful to determine the Viscoelastic Linear Range (VLR), previous to the critical point (dashed line).

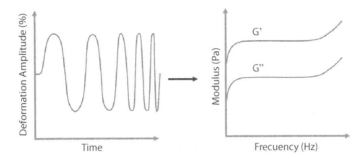

Figure 4.5 Rheological test: Frequency sweep of deformation. The determination is performed under a constant deformation amplitude.

of cross-linking or composition, giving information about the interaction forces between the polymer chains and their nature, allowing, for example, to differentiate covalent interactions from physical ones. Generally, when the material is deformed at high frequencies, the high deformation rate does not allow inelastic relaxation of the material (or the displacement between domains of the polymer) and behaves like a solid (increase in module values is observed). On the contrary, at low frequencies the relaxation of the stressed material can be observed, due to inelastic deformation (a decrease in the values of the modules is observed). This last phenomenon is more frequently observed when the greatest contribution to G' is given by physical interactions, which can relax for a long time, and is less observed when the greatest contribution is covalent, because these unions do not allow the relaxation of the material. In this way, degrees of cross-linking and types of interactions between the measured hydrogels can be compared.

4.4.2 Determination of Viscoelastic Linear Range

Through these rheological tests, fundamental mechanical parameters can be determined to characterize the material used in the actuator and to know the limits of applicability of material developed within the device. The sample to be analyzed may have different geometries, but it is preferable to prepare a disk-shaped specimen thereof (whose diameter depends on the available geometry) in general, samples of 8-, 10-, 20-, or 25-mm diameter are analyzed, by 1–2-mm thick. In this test, the sample is placed between the plates of the rheometer and a strain amplitude scan of 0.01% to 100% is applied at a constant frequency of 1–10 Hz. The commonly used temperature is 20°C, but the linear range of the material at other temperatures can be studied.

4.4.3   Swelling Kinetics

As mentioned earlier, the most important property that hydrogels possess and for which it is widely used in actuators is their swelling capacity. For this reason, in this section, some principles governing this property will be discussed.

The swelling degree of the hydrogels is determined by the system free energy balance, according to the network expansion (dependent on the crosslinking density), the free energy of mixing (relative to the interactions between the polymer chains and solvent molecules), as well as mixing entropy (Flory-Rehner Theory). Considering a constant crosslinking degree, the energy and entropic factors depend mainly on the solvent and polymer interactions and the temperature. Therefore, with a temperature shift or changes in the nature of the solvent (such as chemical composition, ionic force, or pH), the swelling degree of the hydrogels can be affected.

Swelling is a transient process from the non-solvated vitreous or a partial gummy state to a relaxed gummy state. It has been demonstrated that absorption processes of solvent-polymer systems often do not behave according to classical diffusion theory. The slow reorientation of polymer molecules can origin a wide variety of anomalous effects both in permeation and absorption experiments, mainly when assays are performed near or below the glass transition temperature (Tg). According to the diffusional behavior, different categories can be considered. A first case is Fickian transport, usually found when Tg of the polymer is much lower than room temperature. Here, the water easily penetrates the gummy network without major impediments, since polymer chains have great mobility. Therefore, the diffusion rate of the solvent is lower than the polymer relaxation. This conduct is characterized by a linear increase of the hydrogel mass with respect to the square root of the absorption time. In the other hand, non-Fickian diffusion is frequently found when the polymer Tg is above the working temperature and the polymer chains do not have much mobility to allow a rapid penetration of water into the polymer. There are two kinds of non-Fickian behavior, the transport type II and the anomalous diffusion. If diffusion of the solvent takes place much quicker than the relaxation speed of the chains, it is named type II. Thus, the rate of mass change is a direct function of time. An abnormal transport is found when chain relaxation and solvent diffusion rates are similar. To determine the diffusion mechanism in polymer networks, empirical equation (4.2) is used:

$$M_t/M^\infty = kt^n \tag{4.2}$$

Where the constants n and k are characteristic of the solvent-polymer system. The diffusional exponent n is related to the hydrogel shape and interactions with solutes (e.g., a bioactive for drug delivery). Thus, determination of n gives valuable information from the physical mechanism that controls solute uptake. For cylindrical shaped hydrogels, if $0 \leq n \leq 0.5$ diffusion is Fickian, while for $n = 1$, a type 2 mechanism can be expected, and for $n > 0.5$, it would be anomalous.

4.4.4 Experimental Determination of the Swelling Mechanism

One of the properties of hydrogels that must be determined for the development of a hydrogel is the swelling kinetics. Although there are several methodologies, the most appropriate is to take a gel disc and dehydrate until obtaining a polymer in sol state. Subsequently, the dry mass is determined and after being in contact with water at different times. With the results a curve of log Mt/M is constructed based on the log t, in which the slope can be used to calculate the diffusional exponent. In general, the first half of the recorded curve is adjusted. The results obtained allow us to correlate the displacement speed of the actuator with the hydrogel swelling, as well as the water distribution within the material.

4.4.5 Study of the Network Parameters

From the theory of elasticity and the equilibrium theory, the structures and attributes of swollen hydrogels such as the volume fraction of polymer in swollen state, the distance between crosslinking points and the polymer molecular weight between those points can be described. The volume of polymer fraction in a swollen state describes the amount of liquid that may be contained in a hydrogel and is defined as the ratio of the volume of the polymer (Vp) to the volume of the swollen gel (Vs) (equation (4.3)):

$$V_{2,s} = Vp/Vs \tag{4.3}$$

While the distance between intersecting points ζ can be calculated using equation (4.4):

$$\zeta = V_{2,s}^{-1/3} . 1 . (Cn.2Mc/Mr)^{1/2} \tag{4.4}$$

Where Mc is the molecular weight between crosslinks, Mr is the molecular weight of the repetitive units, Cn is the Flory radius, and l is the carbon-carbon bond distance. ζ is an approximation of the distance between polymer chains that define the window for the transport of solvents, ions, and drugs.

4.4.6   Hydrogel Swelling and Porosity of the Network

The main characteristic of hydrogels is its swelling capacity when they hydrate in contact with an aqueous solution. From the dried state, the small molecules attack the surface of the hydrogel and penetrate the polymer network during the first contact with solvent. Here, the gummy region behaves distinctly from the non-solvated vitreous phase of the hydrogel with a restricted movement. Commonly, the mesh of the structure in the rubber phase will begin to expand easing the penetration of new solvent molecules into the network. In equilibrium, when elasticity and osmotic forces are balanced, equilibrium swelling is achieved. In the case of non-charged hydrogels, the polymer-polymer repulsion is originated by Van der Waals forces, but the expansion is not very significant compared to polyelectrolyte gels. In these charged polymers, the applied pressure is given by the osmotic pressure of the counterions, generating a great incorporation of solvent compared to the neutral networks. On the other hand, the swelling rate is determined by several physicochemical parameters, particularly the porosity size and the type of porous structure. Based on these properties, hydrogels can be classified as nonporous, microporous, macroporous, and superporous. Non-porous gels provide limited solute transport by diffusion through free volume. While the pore size is approximated to the size of diffusive solutes, transport takes place by a mixture of molecular diffusion and convection in pores filled with water. Macroporous hydrogels have long pores up to 1 um, thus generally much bigger than the diffusive solutes. Therefore, the effective diffusion coefficient of solute can be represented through the diffusion coefficient of the solute in the pores filled with water. Superporous hydrogels have pores in the range of several hundred micrometers. Most pores within the hydrogels are connected to each other, acting as a capillary system which promotes a fast water uptake. Therefore, it can swell rapidly due to a simple capillary effect rather than absorption.

4.4.7   Experimental Determination of Pore Morphology of a Hydrogel

The pores of a hydrogel are those that define the transport kinetics of species and their swelling capacity; consequently, study their morphology is essential. To observe the morphology of pores, it is not possible to simply dry a hydrogel and observe by microscopy because the collapse of the structure occurs when the solvent is removed. For this reason, to observe the pores of a hydrogel, the structure must be preserved as completely as possible during solvent removal. Before starting, the hydrogel should be maintained in the fully hydration state (controlling with balance its constant mass) and a 1–2 mm diameter fragment is separated. Then, it is taken with tweezers and incorporated into a container with liquid air. In these circumstances, an extremely rapid freezing of the material is produced, so the pores of the material are insignificantly damaged. Note that the fragment to be studied must have small dimensions to ensure that the freezing is fast enough. Subsequently, the frozen sample is taken and placed in a previously cooled freeze-drying equipment (−60°C) and subjected to vacuum to sublimate the water present in the sample. Subsequently, the dried sample is taken very carefully (since its consistency is usually very fragile) and is placed in a conductive stab of the SEM microscope. The sample is usually taken carefully with two tweezers and fractured by traction. Then, it is metallized, in general, with a layer of gold and visualized in the microscope. The best magnifications for observing the structural characteristics of the pores are 100×, 1000×, and 10,000×. In general, larger magnifications require a lot of time, the electron beam alter the sample morphology and do not provide very relevant structural information.

4.5   pH-Sensitive Hydrogels

Several substances are known to alter the ionic equilibrium of water. Changes in the concentration of protons [H+] and hydroxyl [HO−] ions are known as changes in the pH of the solution. As water is the most abundant solvent in our planet, these equilibriums play a fundamental role in nature. Furthermore, water is the main component on living organisms as well as in hydrogels, and alterations on pH often promotes predictable changes on the substances that remain in contact with this solvent.

The most frequent underlying mechanism for pH sensitivity relies on an equilibrium of ionic charges into the hydrogel network. The polymeric chains into the network are endowed with functional groups that can change its electrostatic charge with a change in pH. This behavior is illustrated in Figure 4.6 for hydrogels with acidic functional groups. When the aqueous solution has an acidic pH, virtually, all the acid functional groups in the network remain protonated, thus having a neutral electrostatic charge. However, when pH is raised, the acidic GF tends to

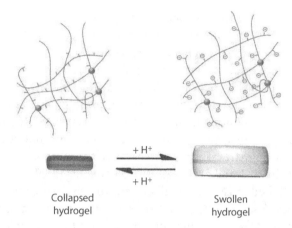

Figure 4.6 pH-responsive hydrogels. The presence of electric charges is responsible for the expansion of the material.

liberate protons (H+) to the aqueous solution and the polymer chains became negatively charged. The presence of multiple charges in the network generates electrostatic repulsion between the polymer chains along with an increased need for solvation to stabilize the charges. Thus, this phenomenon promotes an increase of the pore sizes, major water absorption, and, finally, a hydrogel volume increase. The most used pH sensitive functional groups are carboxylic acids (−COOH), followed by amine derivatives (−NH2) which can be found in natural polymers or obtained in synthetic polymerizations. The selected functional group determines the pH range in which the hydrogel will change its properties. The most used polymer for pH sensitive hydrogels development is poly-(acrylic acid). These hydrogels are frequently applied for products of common use such as diapers or for irrigation and both the polymer and its monomer are commercially available. This has promoted the development of low cost actuators [14].

4.6 Thermosensitive Hydrogels

Thermosensitive hydrogels are the most widely used for the development of actuators; therefore, the properties and general characteristics of this type of materials are mentioned below.

Thermosensitive hydrogels are a type of intelligent materials that are very studied, because the temperature is an easy variable to control both *in vitro* and *in vivo* these are promising in biomedical applications. The thermosensitive polymers that have sensitivity in aqueous solution show a phase change due to changes in the temperature of the solution. It is usually an abrupt change that occurs at a certain temperature, or in a limited range of temperatures. The average temperature of the transition is known as Phase Transition Temperature (PTT). The most commonly used polymers show a homogeneous liquid phase below the PTT, but they separate into a new phase (solid or gel) as the temperature rises above the PTT. In this type of polymers, PTT is also known as a critical lower solubility temperature or LCST since, above this temperature, the polymer precipitates as its solubility decreases. There are also polymers that have the opposite behavior, showing a rapid transition from the solid/gel state to the soluble when temperature is increased (UCST, upper critical solution temperature) [15]. The thermal stimulus is frequently used to cause the gelation of soluble polymers, obtaining a reversible behavior of the sol-gel type, particularly useful in the generation of injectable hydrogels [16]. In the case of crosslinked heat sensitive hydrogels, the degree of swelling of the gel becomes temperature dependent and changes abruptly above and below the PTT of the polymer.

Although there are several thermosensitive polymers, the most extensively studied as actuator has been poly (N-isopropylacrylamide) (p-NIPAm, Figure 4.7) [16]. p-NIPAm is obtained from the N-isopropylacrylamide monomer (NIPAm) and has a completely reversible thermal transition in aqueous media, with an LCST of 32°C. The proximity of its PTT with body temperature (between 36.5°C and 37.5°C) is one of the reasons why this polymer is studied. Furthermore, the NIPAm co-polymerization with other monomers allows to modify the PPT of the synthesized hydrogel, adjusting it to the desired value in the application of interest, for example, reaching values close to 37°C for actuators designed in biomedical applications. In general, co-polymerization of NIPAm with hydrophilic co-monomers generates an increase in PTT while hydrophobic comonomers decrease PTT [17]. The phase transition of this polymer is produced by the presence of hydrophilic functional groups (amide group), together

Figure 4.7 p-NIPAm temperature induced volume phase transition.

with hydrophobic functional groups (isopropyl group and polymer main chain) in its structure. In this system, the balance of hydrophilic/ hydrophobic interactions between the polymer and water is temperature dependent. For example, in a crosslinked p-NIPAm hydrogel, at temperatures below 32°C, hydrophilic interactions between the polymer and water are predominant, and therefore, the hydrogel is at its maximum swelling. When the temperature is high above 32°C, the polymer-solvent hydrophilic interactions weaken and polymer-polymer interactions between the isopropyl groups of the p-NIPAm side chains and their amide groups are favored. Consequently, the polymer chains are approaching and the gel collapses.

However, although the biocompatibility of the homopolymer has been proven, the majority of p-NIPAm copolymers have not been studied *in vivo*, due to the high cost and ethical restrictions of this type of analysis. This is a considerable aspect, considering that the copolymerization of NIPAm is generally necessary for the control of PTT and other material properties.

The use of ionic poly-liquids is an interesting alternative to achieve thermosensitive actuators. These materials can be obtained by the radical reaction of an ionic liquid that contains a vinyl group in the presence of a crosslinking (di-vinyl) molecule and a radical initiator. The material obtained showed an LCST, collapsing due to an increase in temperature; however, unlike p-NIPAm, this property gradually occurs over 20°C to 70°C. With this material, a microfluidic actuator device was constructed and the flow through them was successfully controlled [18].

4.7 Composite Materials Containing Hydrogels

Although there are many actuators based on thermo-sensitive hydrogels, sensitive pH or their combinations such as the hydrogel of p-NIPAM and poly(2-(dimethyl amino) ethyl methacrylate) (p-DMAEMA) developed to obtain simultaneous pH and temperature response [19]. Many hydrogels present problems mainly due to their poor mechanical properties. For this reason, research has been conducted to improve this feature by adding reinforcement materials. An alternative is assayed through layer synthesis and other polymers such as polyvinyl alcohol (PVA) are added to achieve a stronger material [19]. In another case, the incorporation of elastomers and hydrogels in laminates is proposed to improve the properties of hydrogel [20]. While in other cases, the molecule added to the hydrogel is incorporated to give it new responses to stimuli. Such is the case of polyacrylamide and acrylic gels in which spiropyran was immobilized, which gives the actuator sensitivity to light [21]. In this sense, there are many substances described for modifying the response of hydrogels but are two relevant materials that deserve an independent section. Hydrogels modified with carbon family such as graphene or carbon nanotubes and those modified with metals, either in the form of nanoparticles, particles, oxides, or ions in which the remarkable process is ion-printing.

4.8 Ion-Printing

A very important modification that can be made in hydrogels is the controlled incorporation of ions into the hydrogel structure. Through the ion-printing process, ions can be introduced by an electrochemical reaction. For this, one of the most widespread methods is to place the hydrogel between two metal electrodes. One of the electrodes is at negative (reduction) potential, while the other at positive (oxidation). In general, the negative counter electrode is an aluminum foil that covers the whole face of a film-shaped hydrogel. While on the opposite side, an electrode that is generally thin as a wire is placed in contact with the hydrogel, to select the site at which the actuation will occur. Depending on the ion to be deposited, the electrode material (iron for example) is chosen. To generate ions, a continuous potential is generally applied (which is usually 1 to 10 V). The presence of these ions in the hydrogel generate complexes in the hydrogel that change their swelling property which behaves like a new composition in the modified area. Generally, due to this interaction, the swelling of the hydrogel is reduced, and for this reason, the sheet bends towards the site of the deposited ion. Such interaction is reversible, and once the ions are removed, the hydrogel returns to its original state (Figure 4.8) [22].

Using this principle, pH-dependent actuators were developed using the ability of catechol to form complexes with different molecules such as ferric ion. For this, a polymer is prepared from an acrylated catechol, and in the presence of other monomers, a hydrogel is obtained. Then, the process of ion-printing is performed the bending of the film takes place. Thus, the $-OH$ groups of catechol are first protected with boron, and then, the ion-printing process is performed. Iron and aluminum are used as electrodes. Once the potential deoxidation is applied, iron ion (3–10 V) is deposited in certain places where the ion is going to bind to catechol. A saline solution (NaCl) in acidic medium is used to remove borate

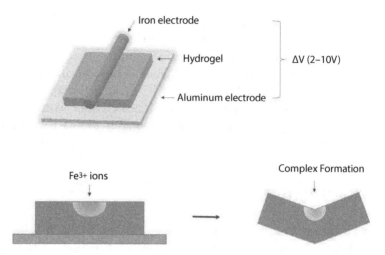

Figure 4.8 Ion-printing process controls the incorporation of ions into the hydrogel structure.

and allow iron to bind. Since the complex can have different number of ligands at different pHs, there is a pH dependent actuator [23]. These ion-printing systems with catechol analogs can be made with different metals, with slight differences between them. It has been found that the actuation angle is proportional to the ion charge, with $Ti^{4+} > Fe^{3+} > Al^{3+} = Cu^{2+} = Zn^{2+}$, a similar profile has the actuation kinetics [24].

4.9 Electroosmotic Effect—Donnan

On the other hand, one way to control the displacement of ions that do not form complexes is by the electroosmotic effect. For this, an ionic hydrogel is metallized on both sides and two electrodes are generated. The ionic polymers commonly used for these devices are fluorinated ion exchange resins. The application of a potential difference produces the displacement of the mobile ions towards one of the electrodes, generating a difference in the chemical potential due to the asymmetric distribution of ions, causing the swelling of the hydrogel. The bending would be produced by a local swelling given by a volume effect of the accumulation of counterions. Water is the commonly used solvent, but if it is desired the actuator to work in air, the water is replaced by ionic liquids. Under these circumstances, there is a gain in actuator stability but its speed decreases. In addition, the performance depends on the size and mobility of the ions. An actuator based on this principle was proposed using chitosan. This polymer has been used because it contains carbonyl groups and amines in its structure. Thus, the electroosmotic effect (Donnan) can be used when an electric field of about 10 V is applied. It has the advantage that it can be printed, and therefore, the geometry control of the material maximize the response in magnitude and action speed [25].

4.10 Graphene Modified Hydrogels

Graphene is a carbon sheet forming a hexagonal structure by covalent sp^2 bonds. It has unique mechanical, electrical, and optical properties [26]. In general, many actuators have been developed incorporating this type of materials, but from the practical point of view, the most interesting are those in which graphene is incorporated throughout a thermosensitive hydrogel such as p-NIPAM. The use of p-NIPAM with graphene oxide has allowed the development of laser-activated optical devices for fluid switching in valve-type actuators capable of responding in a few seconds [27]. Thermosensitive actuators have been obtained from heterogeneous p-NIPAM hydrogels. They were prepared from graphene oxide incorporated into the p-NIPAM hydrogel network. Graphene oxide (GO) was then reduced with a 355-nm laser on one of the faces of the sheet-shaped hydrogel. The resulting material, containing graphene oxide on one side and reduced graphene on the other. Hence, modified p-NIPAM responds differently to temperature in each face. For this reason, when this actuator placed in hot water, the film is deformed, producing bending towards graphene oxide (Figure 4.9) [28].

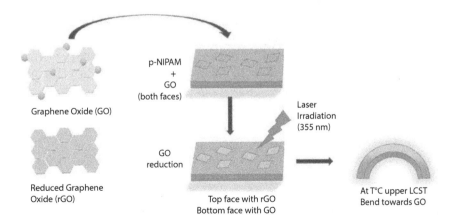

Figure 4.9 Thermo-sensitive hydrogels modified with graphene.

In other studies, it was found that as the amount of GO increases the network pore size decreases, which explains the behavior of the actuator when is modified with graphene [29]. However, to improve the mechanical properties of the material, it is usually reinforced with cellulose and/or nanoparticulate ceramic nanocrystals. Furthermore, graphene oxide can be absorbed in a p-NIPAM hydrogel and 4-hydroxybutyl acrylate, allowing to control the bending area of the hydrogel. In this way, actuators with almost infinite shapes can be achieved, to get complex movements that can be very useful for many applications [30].

4.11   Actuator Geometry

4.11.1   Deformation Behavior According to Hydrogel Moulding

Designed shaping of the hydrogels can be used to promote a predefined movement from an initial to a final shape. While several geometries can be achieved, the most common rods and films together with some others will be briefly discussed in this section.

4.11.1.1   Rods

This type of actuator is obtained with elongated cylindrical structures containing at least two polymeric components with different swelling response against stimuli or only one of them while the other is inert. The idea of operation is based on the background of bimetals. These were developed about a hundred years ago and are constituted by a metal structure made up of two metals with different coefficient of thermal expansion. Hence, temperature changes inevitably cause the bending of the system. The simplest example can be obtained by forming a bilayer rod with a polymer containing carbonyl groups in its structure (such as polyacrylic acid) and one amino group (DMAEMA for example). Exposure of the rod to an acidic medium causes the collapse of the acidic layer and the expansion of the basic one. Under these circumstances, the deformation of the rod towards the side of the polycarboxylic polymer. If the pH increases, a torsion is produced to the other side. To control the curvature of a rod, a close relationship can be found with the swelling of the hydrogel. Furthermore, many developed hydrogels usually have poor mechanical properties that are not very suitable, and it was solved for the researcher adding another inert molecule that perform a reinforcement of hydrogel, such as PVA,

Rod Actuators

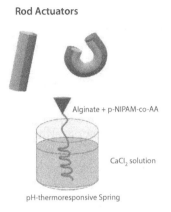

Alginate + p-NIPAM-co-AA

CaCl₂ solution

pH-thermoresponsive Spring

Figure 4.10 Spring actuator obtained with alginate- p-NIPAM-co-AA.

cellulose, etc. In these, it has been shown that this reinforcement significantly improves the mechanical properties of the hydrogel. However, the
effect of the addition of this reinforcement on swelling should be precisely
adjusted, since the interaction between the chains of these interpenetrated
networks is usually very strong and therefore significantly limit the swelling of the hydrogel. A special form of rods are those that are arranged heli-
cally. By this geometry, springs can be formed, which depending on the
structure of the bilayer can produce linear action given by the shortening
or stretching of that spring. The actuators obtained can produce displacement and perform mechanical work. An interesting strategy for obtaining
springs is by mixing linear chains of p-NIPAM-co-AA with alginate, in
which partial mixing is performed and injected into a calcium chloride
solution (Figure 4.10). Upon contact with the saline solution, alginate
forms a gel which collapses adopting a spring shape, which is continuous
while the polymer mixture is injected into the solution. Both polymers
form independent homogeneous phases along the formed tube, obtaining
a bilayer material suitable as thermal actuator [31].

4.11.1.2   Sheets

The most widespread actuators in bibliography are those that belong to
this category. These are based on the principle that the kinetic response of
the actuators depends fundamentally on the thickness of the film, whereby
high response speeds can be obtained if the film is thin. The effect of a
thermosensitive hydrogel associated with a rigid sheet structure and the
behavior of the gel when stimulated has been studied mathematically [32].

However, the most widely developed systems consist of consecutive synthesis or subsequent modification in order to generate two thin layers with different swelling responses (Figure 4.11). A bilayer moisture actuator was developed by combining polydimethylsiloxane (PDMS) with polyacrylamide in a bilayer. The interface is the most complicated area in this system, which was solved by acrylation of the surface of the first, and then photo polymerization of acrylamide on modified PDMS [33]. In these systems, the second layer can be continuous, while in many cases, it is discontinuous. In the same way as described above, the film deforms towards the side of the component that swells less after the application of the stimulus. Although, when the two layers are continuous throughout the entire film, predicting the exact flexion is not easy. Such systems were analyzed with excellent results by Stoychev *et al.* However, it is more easily predictable to build a film of a hydrogel and generate a second layer in specific sites to achieve stimuli sensitive areas capable to get flexion in response to a stimulus. With this strategy, the development of actuators capable of bend at different sites and forming very complex 3D structures and easily predictable through ion-printing has been achieved [25].

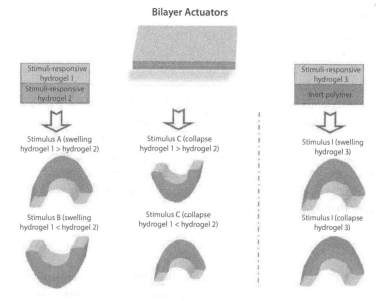

Figure 4.11 Examples of bending in different bilayer actuators.

4.11.1.3   Other Geometries

In these cases, the hydrogels are placed inside a cavity that is partially or totally filled with the hydrogel. There are two possibilities, to place the gel in liquid form containing monomers and initiator, to then polymerize by a light or heat stimulus as mentioned above. In this case, the hydrogel perfectly adopts the shape of the reservoir in which it was placed. On the other hand, the hydrogel can be synthesized in the form of a disc or sphere and placed inside the reservoir. The reservoir where the hydrogel is placed usually has very particular characteristics. In general, this material has a rigid and a flexible part or is completely flexible. Thus, after stimulus, the hydrogel swelling causes at least partial deformation of the reservoir. Several bibliography examples of this reservoir type can be cited, most of them belong to actuators that form valves for fluid control. The circulation of fluids through the valve is generally controlled by the temperature or pH depending on the polymer composition inside the actuator. A valve can be constructed based on two principles: the hydrogel response to the same fluid that controls the valve. In this case, the fluid passage depends on the characteristics of this fluid, for example if the actuator contains a poly-carboxylic hydrogel and the circulating fluid has an acidic pH, the valve is kept open due to the collapse of the hydrogel. However, when the pH of the solution becomes basic, the hydrogel changes its swelling state, increasing its volume, blocking the passage of fluid into the channel formed by the reservoir and cutting off the liquid flow (Figure 4.12). On the other hand, the hydrogel can control the passage of a fluid without having contact with it, in this case another connection is needed because the hydrogel is controlled by an independent fluid flow. For control, the control fluid change

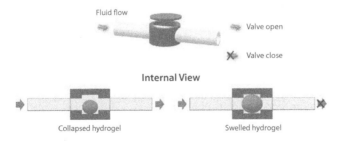

Figure 4.12 Example of a simple actuator for fluid control.

the state of the hydrogel from swollen to collapsed state consecutively, while the passage of the liquid through the other channel is cut or allowed. This system has the advantage that the composition of the liquid to be controlled does not matter, while on the other hand, it requires the design of a slightly more complex actuator. Another approach that is very economical and simple actuators has been designed using gel spheres, glued to an inert film or within an inert tube and solvent permeable. With this type of geometry, actuators can be achieved that can be flexed in a spiral or act in a linear fashion [14].

4.12   Actuators as Fluid Reservoir

As an alternative, in which only the hydrogel is used as a simple reservoir, without taking advantage of its stimulus response characteristics, it was proposed by MIT researchers. In this case, the hydrogel is used as a reservoir of hydraulic fluid. A hydrogel is synthesized forming a stretched container. But when the hydraulic liquid is invaded, the pressure of the liquid causes the expansion of the reservoir generating the action. This actuator was able to respond in less than a second to changes in the liquid pressure inside.

4.13   Actuator Based on Hydrogels for Medical Applications

In general, the actuators are very versatile devices that can be used for a large number of applications. For example, for the control of the release of chemical substances. As the actuator engineered for the controlled release of heparin [34] or the one developed by gelatin-PVA solutions forming a hydrogel that loads drugs such as the anticancer doxorubicin. This last incorporates nanoparticles and can be magnetically guided to the target site and decompose with infrared radiation to release the drug [35]. A great advantage that hydrogel actuators have is that they are highly compatible and can be adapted to complex biological geometries. Another interesting application is as a cochlear implant in the ear. The introduction of a curved implant is very difficult so it must be straight; however once inside the organ, it must take a curved shape. To get around this difficulty, a silicone actuator containing hydrogel particles was made. When it was inserted into the ear, the implant was straight, but once inserted, it was hydrated in the ear fluid and adopted the proper curved shape [36]. On the other hand, miniaturized devices can be obtained, using prototyping techniques such as 3D printing or as humidity sensors that were developed with peg diacrylate, which were polymerized controlled by laser light, which achieved micrometric structures, great control conformational and mainly thanks to its small size, show response to stimuli in a few seconds [37].

4.14   Conclusions and Future Perspectives

Hydrogels are synthetic materials that have very similar characteristics to biological tissues, so they are the most likely to be successfully integrated into living organisms and/or mimic its behavior. Furthermore, the versatility of these materials enables the obtainment of a huge number of smart features such as high elongation capacity, high surface energy density (similar to that of muscle), biodegradability, among others. On the other hand, they are not able to produce a high force like other biological actuators, while the response time is a characteristic that continues today as a challenge to be improved in future developments and scientific studies.

The future of actuators is highly promising. The development of systems helped by 3D printing technologies, modification techniques controlled in very small dimensions (i.e., based on light beams), together with chemical synthesis techniques that are able to expand the variety and compositional control of the hydrogels are envisioned. When the confluence of these technologies occurs, together with professionals working in an interdisciplinary way, better and sophisticated actuators will be achieved and, in the near future, reach those intelligent actuators that behave like robots, which we are hoping arrive at (Figure 4.13).

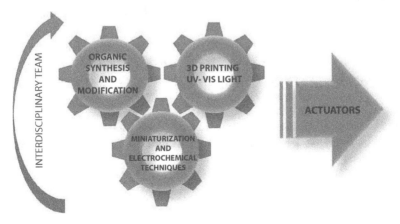

Figure 4.13 Schematic representation of the gears necessary to carry out a successful development of actuators.

Acknowledgments

We would like to thank Prof. Dr. Inamuddin, M. Phil. for his invitation to join us in his book project. Financial support from FCQ-UNC and IPQA - CONICET (Argentina) is gratefully acknowledged.

References

1. Peppas, N.A., Buresa, P., Leobandunga, W., Ichikawa, H., Hydrogels in pharmaceutical formulations. *Eur. J. Pharm. Biopharm.*, 50, 27–46, 2000.
2. Ullah, F., Othman, M.B.H., Javed, F., Ahmad, Z., Akil, H.M., Classification, processing and application of hydrogels: A review. *Mater. Sci. Eng. C*, 57, 414–433, 2015.
3. Guilherme, M.R. *et al.*, Superabsorbent hydrogels based on polysaccharides for application in agriculture as soil conditioner and nutrient carrier: A review. *Eur. Polym. J.*, 72, 365–385, 2015.
4. Buenger, D., Topuz, F., Groll, J., Hydrogels in sensing applications. *Prog. Polym. Sci.*, 37, 12, 1678–1719, 2012.
5. Romero, M.R., Wolfel, A., Igarzabal, C.I.A., Smart valve: Polymer actuator to moisture soil control. *Sens. Actuators B Chem.*, 234, 53–62, 2016.
6. Mahinroosta, M., Jomeh Farsangi, Z., Allahverdi, A., Shakoori, Z., Hydrogels as intelligent materials: A brief review of synthesis, properties and applications. *Mater. Today Chem.*, 8, 42–55, 2018.
7. Kamila, S., Introduction, Classification and Applications of Smart Materials: An Overview. *Am. J. Appl. Sci.*, 10, 8, 876–880, Aug. 2013.
8. P., A., M., S.K., Sharma, S., *Hydrogels From simple networks to smart materials—Advances and applications*, Elsevier Inc, 2018.
9. Willner, I., Stimuli-Controlled Hydrogels and Their Applications. *Acc. Chem. Res.*, 50, 4, 657–658, 2017.
10. Ito, S. *et al.*, Effects of resin hydrophilicity on water sorption and changes in modulus of elasticity. *Biomaterials*, 26, 33, 6449–6459, 2005.
11. Abdurrahmanoglu, S., Can, V., Okay, O., Design of high-toughness polyacrylamide hydrogels by hydrophobic modification. *Polymer (Guildf)*, 50, 23, 5449–5455, 2009.
12. Guo, X.Q., Qiu, K.Y., De Feng, X., Studies on the kinetics and initiation mechanism of acrylamide polymerization using the persulfate/aliphatic diamine system as initiator. *Macromol. Chem. Phys.*, 191, 577–587, 1990.
13. Vedadghavami, A. *et al.*, Manufacturing of hydrogel biomaterials with controlled mechanical properties for tissue engineering applications. *Acta Biomater.*, 62, 42–63, 2017.

14. Velders, A.H., Dijksman, J.A., Saggiomo, V., Hydrogel Actuators as Responsive Instruments for Cheap Open Technology (HARICOT). *Appl. Mater. Today*, 9, 271–275, 2017.
15. Gong, C. *et al.*, Thermosensitive Polymeric Hydrogels As Drug Delivery Systems. *Curr. Med. Chem.*, 20, 1, 79–94, 2012.
16. Jeong, B., Kim, S.W., Bae, Y.H., Thermosensitive sol-gel reversible hydrogels. *Adv. Drug Deliv. Rev.*, 64, SUPPL., 154–162, Dec. 2012.
17. Lanzalaco, S. and Armelin, E., Poly(N-isopropylacrylamide) and Copolymers: A Review on Recent Progresses in Biomedical Applications. *Gels*, 3, 4, 36, 2017.
18. Wu, S., Yu, F., Dong, H., Cao, X., A hydrogel actuator with flexible folding deformation and shape programming via using sodium carboxymethyl cellulose and acrylic acid. *Carbohydr. Polym.*, 173, 526–534, 2017.
19. Cheng, Y., Ren, K., Yang, D., Wei, J., Bilayer-type fluorescence hydrogels with intelligent response serve as temperature/pH driven soft actuators. *Sens. Actuators B Chem.*, 255, 3117–3126, 2018.
20. Hubbard, A.M. *et al.*, Hydrogel/Elastomer Laminates Bonded via Fabric Interphases for Stimuli-Responsive Actuators. *Matter*, 1, 1–16, 2019.
21. Francis, W., Dunne, A., Delaney, C., Florea, L., Diamond, D., Spiropyran based hydrogels actuators—Walking in the light. *Sens. Actuators B Chem.*, 250, 608–616, 2017.
22. Baker, A.B., Wass, D.F., Trask, R.S., Thermally induced reversible and reprogrammable actuation of tough hydrogels utilising ionoprinting and iron coordination chemistry. *Sens. Actuators B Chem.*, 254, 519–525, 2018.
23. Lee, B.P., Lin, M.H., Narkar, A., Konst, S., Wilharm, R., Modulating the movement of hydrogel actuator based on catechol-iron ion coordination chemistry. *Sens. Actuators B Chem.*, 206, 456–462, 2015.
24. Lee, B.P., Narkar, A., Wilharm, R., Effect of metal ion type on the movement of hydrogel actuator based on catechol-metal ion coordination chemistry. *Sens. Actuators B Chem.*, 227, 248–254, 2016.
25. Zolfagharian, A. *et al.*, Development and analysis of a 3D printed hydrogel soft actuator. *Sens. Actuators A*, 265, 94–101, 2017.
26. Novoselov, K.S., Fal'ko, V.I., Colombo, L., Gellert, P.R., Schwab, M.G., Kim, K., A roadmap for graphene. *Nature*, 490, 192, 2012.
27. Breuer, L., Pilas, J., Guthmann, E., Schöning, M.J., Thoelen, R., Wagner, T., Towards light-addressable flow control: Responsive hydrogels with incorporated graphene oxide as laser-driven actuator structures within microfluidic channels. *Sens. Actuators B Chem.*, 288, March, 579–585, 2019.
28. Guo, L. *et al.*, Fast fabrication of graphene oxide/reduced graphene oxide hybrid hydrogels for thermosensitive smart actuator utilizing laser irradiation. *Mater. Lett.*, 237, 245–248, 2019.

29. Zhao, Q., Liang, Y., Ren, L., Qiu, F., Zhang, Z., Ren, L., Study on temperature and near-infrared driving characteristics of hydrogel actuator fabricated via molding and 3D printing. *J. Mech. Behav. Biomed. Mater.*, 78, September 2017, 395–403, 2018.

30. Zhao, Q., Liang, Y., Ren, L., Yu, Z., Zhang, Z., Ren, L., Bionic intelligent hydrogel actuators with multimodal deformation and locomotion. *Nano Energy*, 51, June, 621–631, 2018.

31. Yoshida, K., Nakajima, S., Kawano, R., Ono, H., Spring-shaped stimuli-responsive hydrogel actuator with large deformation. *Sens. Actuators: B. Chem.*, 272, 361–368, 2018.

32. Hu, Y., Kim, P., Aizenberg, J., Harnessing structural instability and material instability in the hydrogel-actuated integrated responsive structures (HAIRS). *Extreme Mech. Lett.*, 13, 84–90, 2017.

33. Liu, S., Boatti, E., Bertoldi, K., Kramer-Bottiglio, R., Stimuli-induced bi-directional hydrogel unimorph actuators. *Extreme Mech. Lett.*, 21, 35–43, 2018.

34. Guo, H., Dai, W., Miao, Y., Wang, Y., Ma, D., Xue, W., Sustained heparin release actuator achieved from thermal and water activated shape memory hydrogels containing main-chain LC units. *Chem. Eng. J.*, 339, January, 459–467, 2018.

35. Kim, D.-i., Lee, H., Kwon, S.-h., Choi, H., Park, S., Magnetic nano-particles retrievable biodegradable hydrogel microrobot. *Sens. Actuators B Chem.*, 289, March, 65–77, 2019.

36. Stieghorst, J., Tran, B.N., Hadeler, S., Beckmann, D., Doll, T., Hydrogel-Based Actuation for Modiolar Hugging Cochlear Implant Electrode Arrays. *Procedia Eng.*, 168, 1529–1532, 2016.

37. Lv, C. *et al.*, Humidity-responsive actuation of programmable hydrogel microstructures based on 3D printing. *Sens. Actuators B Chem.*, 259, 736–744, 2018.

5

Smart Polymer-Based Chemical Sensors

Dnyandeo Pawar[1,2]*, Ch. N. Rao[1] and Peijiang Cao[1]

[1]College of Materials Science and Engineering, Guangdong Research Center for Interfacial Engineering of Functional Materials, Shenzhen University, Shenzhen, PR China
[2]Key Laboratory of Optoelectronic Devices and Systems of Ministry of Education and Guangdong Province, College of Optoelectronic Engineering, Shenzhen University, Shenzhen, PR China

Abstract

Remarkable development has been done in the field of polymer due to its availability of various synthesis routes, new surface modification methods, and advanced fabrication techniques. In recent past, polymer-based sensors have gained tremendous attention due to involvement of nanotechnology, and therefore, polymer nanocomposite-based sensors changed the complete scenario of chemical sensing owing to their many fascinating properties such as high sensitivity, selectivity, low response/recovery time, and ability to work in a range of temperature. This chapter introduces the study of numerous kinds of polymers in the field of chemical sensors including gas, vapors, humidity, pH, ions, and other chemicals along with their advantages and limitations. The first section focused on the role of polymer and importance in chemical sensing. The second section addressed the different immobilization techniques for the polymer-based chemical sensor development along with their advantages and disadvantages. The third section summarizes various approaches utilized for chemical species detection. The sensors for each chemical parameter detection are discussed in the fourth section. Their sensor performance in tabulated form is also been presented. Numerous techniques with their merits and demerits are also been elaborately discussed. The last section focused on the future perspective of various polymer-based chemical sensors.

Keywords: Conjugated polymers, immobilization, polymeric coatings, polymer composite, polymer sensors

*Corresponding author: pawar.dnyandeo@gmail.com

Inamuddin, Rajender Boddula and Abdullah M. Asiri (eds.) Actuators: Fundamentals, Principles, Materials and Applications, (75–138) © 2020 Scrivener Publishing LLC

5.1 Introduction

Polymer has rehabilitated our day-to-day life beyond our imagination, and now, it became an essential for our continuous and sustainable growth. Due to its tremendous availability of numerous chemical structures, various amazing properties, low cost, easy fabrication techniques, and ability of recycling makes it one of the widely studied classes of material for fabrication of superior chemical sensor. Generally, in sensing domain, sensitivity, selectivity, fast response, and low cost are the key factors for the rapid development of chemical sensor for numerous applications. Polymer-based chemistry is very much familiar and well adopted to sensory field. With advancements of new versatile techniques and methods, it's been possible to manufacture into different shapes like wire, spherical, microfibers, etc. The polymeric material could be easily available or could be synthesized in any form like gels, solutions, self-assembled nanoparticles, suspended solutions, films, or solids [1–4]. Currently, polymer-based composites have attracted great attention of researchers to develop very sensitive, cost-effective, and reliable sensing platform. The polymer composites provide a better mechanical stability, uniform dispersion that leads to enhance in sensitivity and selectivity, and ease for functionalization. Under the exposure of external stimuli such as gas, vapor, ion, or humidity, the chemical properties of polymeric sensing material changes reversibly or irreversibly [5, 6]. In chemical sensing, polymer-based material could be used as a sensing material for target molecule detection, acts as a supporting matrix to give mechanical stability, and facilitates uniform mixing which enhances the sensitivity further. The unique advantages of polymer-based materials include that the surface chemistry can be tuned very easily, flexibility, biocompatible, and immune to degradation.

There are certain requirements of polymer material to be fulfilled to realize a novel chemical sensor. The polymer material must be mechanically, physically, and chemically stable to achieve good functioning lifetime. Moreover, no migration or reorientation of doped material in polymer matrix should occur. The polymer material must be thermally stable at high temperature [2, 7–10]. In chemical sensing, polymer should be stable in humidity environment (depends upon the application), against ambient light, non-toxic, and biocompatible. Polymer must be optically active and transparent in the broad wavelength spectrum in which the sensing being carried out. Thus, these are the basic requirements of any polymer-based sensor for chemical detection.

Over the past decade, the conjugated or conductive polymers (CPs)-based research field is emerged as multidisciplinary approach for sensing numerous chemical moieties in numerous fields including medicine, health care, agriculture, environment [2, 10, 11]. A CPs is defined as a "macromolecule in which the entire main chain is bound together with saturated sigma bonds, and continuous pattern of unsaturated conjugated pi bonds which give rise to an insulating or semiconductor material due to the band gap" [2, 5]. If CPs doped with certain materials, it shows conducting behavior. Therefore, CPs create an extra degree of freedom to tune its optical and optoelectronic properties. Due to these unique properties of CPs, it has been widely explored for the detection of trace amount of chemical or biomolecule, explosives, and toxic content. Figure 5.1 shows the most common structures of CPs those are extensively utilized for the fabrication of chemical sensor.

Considering the scope of polymers, various polymer-based chemical sensors have been fabricated by using numerous techniques and approaches including fiber optics [12–14], chemiresistive [15–17], voltammetry [18], fluorescence [19, 20], and capacitive [21, 22], etc. Since last decade, many review articles have been published and discussed about the fabrication of polymers, utilization of techniques, applications, and future perspective [3–6, 8–10, 23–38].

This chapter is structured in different parts containing the information regarding development of polymer-based gas sensor, vapor sensor, ion sensor, pH sensor, and humidity sensor. The section also consists of advantages and disadvantages of these developed sensors. The last section is emphasized on conclusion and the future perspective of the polymer-based chemical sensors.

5.2 Immobilization Strategies for the Development of Polymer-Based Sensors

The innovative sensing coating plays an important role in chemical sensing. Polymer affinity to certain analyte can be modified by using different immobilization procedures. Numerous immobilization techniques such as adsorption, covalent bonding, and trapping into the polymer matrix have been widely explored for sensor development. The procedures are described as below.

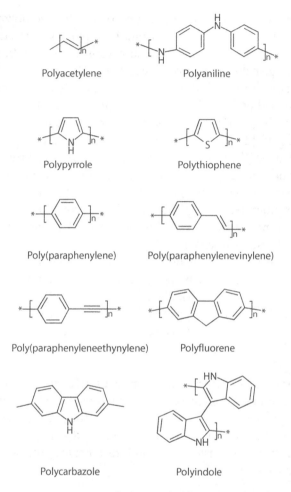

Figure 5.1 Most common structures of conjugated polymers. Reprinted with permission from Ref. [5]. Copyright of Elsevier.

5.2.1 Langmuir-Blodgett (LB) Technique

In LB technique, the layer-by-layer successive deposition can be done by using layer-by-layer of one or more amphiphilic molecules between water and air interface. Both the layers are attached due to electrostatic force and weak force (Van der Waals force). By maintaining constant pressure, the amphiphilic molecule compressed which forms a monolayer [5, 38]. The change in surface pressure can be monitored by electromechanically. This is one of the fine techniques to obtain the film thickness in tens of nanometer per layer. To form a LB film, the molecule either have polymeric or non-polymeric in nature in which it must be spread/dispersed at the surface or adsorbed at the surface positively. This technique is being widely utilized for many polymer-based chemical sensor developments [39–42].

5.2.2 Layer-by-Layer (LbL) Self-Assembled Technique

The LBL is widely adopted method for sensor development due to controlled parameters, designs, and composition. The mechanism of LBL is based on electronic self-assembly in which the two opposite charge polyelectrolytes are get deposited on the substrate [4, 43]. This method is also allowed to include nanoparticles with biological receptors. The LBL growth can be done by using various methods such as spin, immersive, spray, and electromagnetic and fluidic technologies [44]. This method possesses unique advantages such as fabrication of multifunctional film surface without use of any substrate, and could provide strong interaction. However, the main drawback of this technique is that it may lead to destabilization due to the variation of ionic strength. This technique has been used for many chemical parameters detection [44–49].

5.2.3 Covalent Binding

This is the strongest immobilization due to the attachment of covalent bond between receptor and polymer. Most of the functional groups like amino, carboxyl, and hydroxyl groups have been used for this kind of immobilization. Certain polymers like polypyrrole and polythionine possesses amines in their structure which helps in immobilization [5]. Covalent immobilization has certain advantages over electrostatic and mechanical-based methods as the bonding between the indicators is strong, therefore difficult to wash-down quickly and hence provides long term stability [50]. Certain covalent immobilized-based sensors are explored for different chemical detection [51, 52].

5.2.4 Affinity-Based Binding

The selectivity issue can be solved by using this method. This method is based on biomolecular interactions. In this method, the enzymes are immobilized by using biotin and avidin in which one end is attached to the polymer and other end to the receptor [5]. This technique gives enough chemical immobilization and generates more homogeneous structure. This is one of the easy and suitable way applied for immobilization and therefore generally considered for numerous chemical and biosensor fabrication [53–55].

5.3   Approaches for Chemical Detection

Till date, several techniques have been fabricated and demonstrated successfully for the detection of several chemical parameters. Here, we summarized only most popular methods which are extensively explored for detection.

5.3.1   Surface Acoustic Wave (SAW)

In this method, piezoelectric material is commonly used for the detection purpose. A sensitive layer is coated onto the piezoelectric material (mostly used quartz/lithium niobate/lithium tantalate). If any chemical moieties adsorb on to the sensitive layer, oscillating resonant frequency of the material deflects or changes, and hence, the respective amount of analyte can be monitored from the shift generated. In recent past, most of the sensors are projected by using this mechanism [56–58].

5.3.2   Quartz Crystal Microbalance (QCM)

This is widely used for real-time measurement of very small mass change and almost capable to measure less than a nanogram. The sensing layer is deposited on to the quartz surface and measure the frequency shift due to adsorption of any mass. However, modification in surface coating and resonator characteristic is very essential aspect of this technique. Considerable efforts have been made for the development of QCM-based sensors [59–61].

5.3.3   Chemiresistor

This is extensively adopted technique for chemicals detection due to ease in fabrication, sensitivity, and cost-effectiveness. The electrical resistance changes due to adsorption of any chemical species. The sensitive layer is coated on the substrate in between two conducting electrodes which output modifies with respect to external stimuli. The electrical conductivity of the geometry can be easily modulated by doping or by using functionalization technique. Therefore, via this method, most of the CPs are extensively explored for chemical sensor development [40, 62–65].

5.3.4   Optical Approach

Several mechanisms such as fluorescence, photoluminescence, colorimetric, SPR, and interferometric-based techniques are adopted for the development of chemical sensors. The change in light intensity is the function of concentration of analyte. Fluorescence mechanism has been broadly employed for ions detection [66–69]. Some other principles are also been exploited for additional chemical parameter detection [20, 70–77].

5.4   Polymer for Detection of Various Chemical Moieties

In recent years, polymer-based sensors have been broadly applied for sensing various chemicals. In presence of chemical moieties, the physico-chemical properties of polymer get modified which further transform the signal in terms of absorbance, current, acoustic, refractive index, or mass, etc., depending upon the configuration used. Polymers can be tailored in such way that it could detect a single stimulus or can also made to respond multiple stimuli, e.g., gas, vapors, ions, pH, and humidity. Most of the polymers are reversible upon removal of the stimulus. Most of the polymers are conducting and some are possessing swelling ability in presence of chemicals such as gases, vapors, ions, etc. These properties of polymer make it very attractive platform for chemical sensing.

Recently, polymers have been found great interest for gases and vapor sensing. Polymers such as polypyrrole (PPy) nanoparticles (NPs) were used for detection of ammonia. In this work, the urchin like PPy was fabricated by using a dual-nozzle electrospray along with the vapor deposition polymerization technique. The minimum detection limit of this sensor towards NH_3 was approximately 0.01 ppm [78]. The PVA-In_2O_3 (1–5 wt% In_2O_3 loading) nanocomposites were utilized for monitoring NH_3, H_2S, CH_3, CO, and NO in 500 ppb to 100 ppm. The fabricated device showed good responsivity towards H_2S gas but took much higher time to recover in ~410 s [79]. Polymer composite PPy and polyaniline (PANI) prepared by impregnated-oxidation technique has been utilized for acetone gas detection. The maximum sensitivity and minimum response time were observed for PANI with LOD of around 29 ppm [80]. Ionic liquid-based CO_2 sensor has been employed by Mineo *et al.* [81]. The sensing material has great affinity towards CO_2 gas and explored by means quartz crystal microbalance (QCM) in the CO_2 concentration ranging from 17% to 100%. Kukla et al. [82], fabricated thin films by using various polymers such as PANI, PPy, and poly-3-methylthiophene doped with acid sensors array for different volatile organic chemical recognition based on conductivity principle.

The polymer has also been explored for ion sensing. A disposable and simple color-based nylon membrane with 35 sensing areas has been used to identify 13 metal ions with great accuracy [83]. Recently, N^2,N^2,N^9,N^9-tetrabutyl-1,10-phenanthroline-2,9-dicarboxamide (PDAM) acts as a ligand sensitive towards Cd^{2+} ion in aqueous solution by using Raman spectroscopy [84]. The proposed geometry could detect Cd^{2+} in 0–10 mM range. The 2,6-Substituted pyridine derivative-containing conjugated polymers have been established for Pd^{2+} ion based on fluorescence quenching with observed LOD below 1 ppm [85]. Copper ion detection sensor based on colorimetric principle has been proposed by Ding et al. [86]. The PANI/PA-6 nano-fiber/net films were prepared by using electro-spinning/netting process. The sensor could detect 1 ppb to 100 ppm with a LOD 1 ppb.

Certain polymers have great affinity towards pH sensing. The nanoprobe based on $NaYF4:Yb^{3+}, Er^{3+}$ particles modified with polyethylenimine (PEI) attached to pHrodo Red as the fluorescent molecular pH probe has been proposed by Nareoja et al. [87]. It exhibits linear change in the range between pH 5 to pH 7. The poly(3,4-ethylenedioxythiophene (PEDOT) mixed with pH dyes (i.e., Bromothymol Blue and Methyl Orange) have been projected for pH monitoring [88]. It showed well linearity in pH 1–9. Another polymeric materials Polydiacetylenes (PDA), poly(ethylene oxide)-Polydiacetylenes (PEO-PDA), and polyurethane-Polydiacetylenes (PU-PDA) nanofiber mats were employed for pH monitoring from pH 0 to 14 [75]. It was observed that the color of the dye was changed in the range of pH 11–13. Recently, fiber-optic end coated with 5(6)-carboxyfluorescein–PdTFPP sensor has been proposed for pH measurement [89]. The fluorescence-based probe was very sensitive and accurate to external change in the pH varied from 6 to 8.

Polymers have also played major role in humidity sensing. Recently, based on resistive principle polypyrrole/polyoxometalate composite has been used for humidity sensing (11%–98%) with response/recovery time of 1.9 s and 1.1 s, respectively [90]. The cross-linked (PMDS and PPDS) polymer was explored in humidity sensing for human respiration monitoring [91]. It replied quickly in 0.29 s and recovered in 0.47 s during humidity exposure from 33% to 95%. Conducting polymer such as PEDOT:PSS explored to humidity level from 0% to 100% at different impedance value with obtained sensitivity in the RH <1% range based on resistive type [92].

Polymer is widely utilized for sensing numerous chemical parameter monitoring. The details of polymer-based sensor in numerous fields have been provided below.

5.4.1   Polymer-Based Sensors for Gases Detection

Nowadays, tremendous pollution is increasing day by day due to vast use of vehicles and rapid development of industries, and now, it becomes a serious environmental concern. There is immense need to develop a sensor which could detect these trace gases. Generally, metal oxides are mostly explored for gas sensing applications due to its ability of fast charge transfer between the junction, high surface area, and simple synthesis methods. However, these sensors have certain limitations and mainly based on the physico-chemical properties of metal oxides [93, 94]. Certain gases could not be detected or shows minor response by the metal oxides and metals, and therefore, it is required to explore some other materials like polymers [95, 96]. Table 5.1 shows some polymeric-based sensors employed for various gases detection.

Polymer is best choice for gas sensing applications due to their superior properties such as ability to work at room temperature, offer large sensitivity, fast response time, and substrate flexibility with uniformity. Several polymer-based gas sensors have been proposed and fabricated for NH_3, H_2S, CO, CO_2, and CH_4, etc. Depending on the change in physical properties, the polymer-based gas sensors are classified into two types such as conducting (also called CPs) and non-conducting polymers. In CPs, the operation is based on change in electrical resistance of the polymers in presence of gas concentration (shown in Figure 5.1). There are certain polymers such as polyaniline, polypyrrole, and polythiophene, and their derivatives have been extensively used for gas sensing application [5, 97, 98]. However, it should be mentioned that the conductivity of these polymers is less, and therefore, through doping of metal and metal oxides, the sensitivity could be changed significantly [94, 99–104]. In non-conducting polymers, the polymer acts as coating on different devices in which depending upon the external stimuli, the polymer layer causing changes in resonance frequency, refractive index, and other physical properties. The various sensors have been fabricated by using this kind of polymer and technologies [9, 97, 105–109]. The details of various polymer-based gas sensors have been described below.

The interaction of NH_3 with polymer nanocomposite has been studied by Li et al. [101]. The SnO_2 nanosheet is synthesized by using hydrothermal technique and SnO_2/polypyrrole is obtained by using vapor phase

Table 5.1 Polymer-based sensor for various gases detection.

Materials	Type of sensor	Analyte	Detection range	Sensitivity (S)	Limit of Detection (LOD)	Response/ recovery time	Ref.
SnO_2/polypyrrole nanosheet	Resistance based	NH_3	1–200 ppm	~6.2%/ppm	257 ppb	259 s / 468 s	[101]
PPy/TiO_2	QCM	NH_3	10–200 ppm	–	–	100 s / 200 s	[46]
p-toluene sulfonate hexahydrate (PTS)-polyaniline (PAni)	Resistance based	NH_3	5%–200%	225%	5 ppm	112 s/ -	[116]
Gold nanostar-polyaniline composite	Resistance based	NH_3	20–250 ppm	52%	–	15 s/-	[117]
Polymer [(TabH)(AgBr2)] n (1)	Resistance based	NH_3	30–3,000 ppm	197	0.05 ppm	-/-	[118]
Aniline-Tetra-β-carboxyphthalocyanine cobalt(II) [PANI- 2.5TcPcCo]	Resistance based	NH_3	50 ppb to 250 ppm	802.7% @ 100 ppm	10 ppb	~17 s / 5 min.	[119]
Fluorinated difluorobenzothiadiazole-dithienosilole polymer (PDFDT)	Organic thin film transistor	NH_3	1–10 ppm	up to 56%	<1 ppm	-/-	[120]

(Continued)

Table 5.1 Polymer-based sensor for various gases detection. (*Continued*)

Materials	Type of sensor	Analyte	Detection range	Sensitivity (S)	Limit of Detection (LOD)	Response/ recovery time	Ref.
Poly(2,5-bis(3-tetradecylthiophen-2-yl)thieno[3,2-b]thiophene) (PBTTT)	OFET based	NH$_3$	10–100 ppm	12	–	–/10 s	[121]
DPP-bithiophene conjugated polymer pDPPBu-BT	FET based	NH$_3$	0–1,000 ppm	–	< 10 ppb (v/v)	5 s/-	[122]
Solarmer 0.9 wt% in chlorobenzene (CT) -Poly(3-hexylthiophene-2,5-diyl) (P3HT)	Current based	NH$_3$	100–5,600 ppb	–50.3%	100 ppb	–	[111]
Polyaniline-HCl	SPR based	NH$_3$	80–512 ppm	–	0.4 ppm	<90 s/-	[123]
Poly(3-hexylthiophene) (P3HT) Organic	OFET based	NH$_3$	0.01–25 ppm	11.8	8 ppb	–	[110]
TiO$_2$/polyacrilic acid (PAA)	QCM	NH$_3$	0.3–15 ppm	0.17×106 Hz/ mol	0.1 ppm	5 s/-	[61]
PDMS-PMMA	Fabry-Perot optical fiber	NH$_3$	5–500 ppm	4.16 pm/ppm	4.8 ppm	50s/10s	[124]

(Continued)

Table 5.1 Polymer-based sensor for various gases detection. (*Continued*)

Materials	Type of sensor	Analyte	Detection range	Sensitivity (S)	Limit of Detection (LOD)	Response/ recovery time	Ref.
Polyaniline nano-capsules/ zinc oxide hexagonal microdiscs (PANI/ZnO)	Resistance based	H_2S	0.1–100 ppm	40.5% @ 50 ppm	100 ppb	63 s/-	[112]
Cu^{2+}-Doped SnO_2 Nanograin/ Polypyrrole Nanospheres	Current based	H_2S	0.6–50 ppm	1.47	0.05 ppm	-/-	[125]
ZnO-PMMA nanocomposite	Fabry-Perot optical fiber	H_2S	1–5 ppm	Shift = 1.95 nm @ 5 ppm	–	-/27s,	[126]
Ethylenediamine-modified reduced graphene oxide (RGO) and polythiophene (PTh)	Resistance based	NO_2	1–10 ppm	26.36	0.52 ppm	-/-	[127]
Poly(3-hexylthio-phene-2,5-diyl) (P3HT) and poly(9-vinylcarbazole) (PVK) blend	OFET-based	NO_2	0.6–30 ppm	>20,000% for 30 ppm	≈300 ppb	-/-	[113]

(*Continued*)

Table 5.1 Polymer-based sensor for various gases detection. (*Continued*)

Materials	Type of sensor	Analyte	Detection range	Sensitivity (S)	Limit of Detection (LOD)	Response/recovery time	Ref.
Poly(bisdodecyl-quaterthiophene) and poly-(bisdodecylthio-quaterthiophene) (PQT12 and PQTS12)	OFET based	NO_2	1 ppm, 5 ppm	1.7@ 5 ppm	<1 ppm	-/-	[128]
Reduced graphene oxide (rGO)- poly(vinyl alcohol) (PVA)- poly(ether imide) (PEI)	Current based	NO_2	150 ppb–5 ppm	1.03 ppm−1	150 ppb	4 min/10 min	[129]
Poly(3,4-ethylenedioxy-thiophene) (PEDOT)- reduced graphene oxide (RGO)	Resistance based	NO_2	500 ppb–20 ppm	41.7%	–	170– 180 s/70 s	[40]
Poly(methyl methacrylate) (PMMA)- palladium (Pd) nanoparticle (NP)/single-layer graphene (SLG)	Resistance based	H_2	0.025% 2%	66.37%	–	1.81 min/5.52 min	[115]

(*Continued*)

Table 5.1 Polymer-based sensor for various gases detection. (*Continued*)

Materials	Type of sensor	Analyte	Detection range	Sensitivity (S)	Limit of Detection (LOD)	Response/ recovery time	Ref.
Graphene/polyaniline (PANI)	Resistance based	H_2	0.06%–1%	16.57% toward 1% H2	–	-/-	[130]
Copolymer poly(vinyl chloride)-g-poly(oxyethylene methacrylate) (PVC- g-POEM) - mesoporous SnO_2	Resistance based	H_2	3%	–	60 ppm	~10 s/~54 s	[131]
Polypyrrole Nanowires	Resistance based	H_2	600–10,000 ppm	3.88	12 ppm	72 s/-	[114]
Methylated poly(ethylenimine)	Capacitive based	CO_2 and SO_2	0.16%–1.6%, 1%–5%	~8 Hz/ppm, ~20 Hz/ppm	0.011vol %, 0.704vol %	–	[132]
Polyethylenimine (PEI)	Absorption based	CO_2	0–2746 ppm	0.0537%/ppm	20 ppm	-/-	[133]
Poly(N-(3-amidino)-aniline) (PNAAN), coated gold NPs (AuNPs)	Colorimetric based	CO_2	0.0132 – 0.1584 hPa		0.0024 hPa	–	[134]
P(4VP–VBAz)–SWCNTs	Current based	CO_2	0%–2%	~34%	0.03%	200 s/-	[135]

(*Continued*)

Table 5.1 Polymer-based sensor for various gases detection. (*Continued*)

Materials	Type of sensor	Analyte	Detection range	Sensitivity (S)	Limit of Detection (LOD)	Response/ recovery time	Ref.
$^{13}C_{70}$ Fullerene in polystyrene (PS), ethyl cellulose (EC) and an organically modified silica gel ("ormosil"; OS)	Fluorescenc based	O_2	0–150 ppmv	–	EC ~250 ppbv, OS ~320 ppbv, PS ~530 ppbv at 25°C @ 1%	28–43 ms /-	[136]
ZnO/SU-8 hybrid nanocomposite	QCM based	O_2	–	96%	10 mTorr	–	[137]
AgBF4/polyvinylpyrrolidone	QCM	Ethylene	1–7 ppm	51 Hz/ppm	420 ppb	-/-	[138]
Nafion	Cyclic voltammetry based (CV)	CH_4	10%, 50% in N2	–	–	13 s /-	[18]
Polypyrrole–Zinc Oxide (PPy/ZnO)	Resistance based	LPG	1,000–1,800 ppm	32.5%	–	-/ 40 min.	[139]

polymerization process. The polymer composite resistance changes in presence of NH_3 gas. The nanocomposite exhibits good Ohmic contact, and therefore, it decreases the contact resistance which further promote the charge transfer in the interactions between the composite and NH_3. The deprotonation of PPy occurs in presence of NH_3 which further increase the resistance of the sensor and hence increases the magnitude of the sensitivity. It showed remarkable response to NH_3 ~6.2%/ppm under the exposure of 1–200 ppm. However, it has certain limitations such as high response/recovery time of around 259 s/468 s, respectively.

There are a number of OFET-based structures that have been proposed for several gases detection. Recently, Mun et al. [110] have proposed a OFET-based NH_3 sensor by using P3HT as an organic semiconducting layer (OSC) in the concentration range from 0.01 to 25 ppm. Due to large surface area of P3HT, it allows NH_3 to interact and hence the fast charge transfer occurs in between OSC layer and NH_3 gas. It possesses sensitivity of 11.8 with LOD of 8 ppb. The novelty of sensor consists of very low achieved LOD and immune to relative humidity in the ratio of 45%–100%. The breath NH_3 gas sensor has been demonstrated by Yu et al. [111]. Figure 5.2a shows fabrication steps of proposed sensor. The PVP layer is deposited over the top of the ITO along with aluminum electrode followed by the sensing layer. Figure 5.1b shows the scanning electron microscope image of PBDTTT-C-T DL sensor. Figures 5.2c and d show the experimental setup of NH_3 sensing. The sensing response of the sensor is depicted in Figures 5.2e–h. The sensor response in terms of current characteristics for CT-SL and CT/P3HT-DL with different concentrations of NH_3 is shown in Figures 5.2e and f. The NH_3 concentration is varied from 500 ppb to 5,000 ppb. The sensing mechanism is based on redox reaction (shown in Figure 5.2g). The CT acts as a p-type organic material which receives the electron from NH_3, which further decrease the hole concentration of CT, and therefore, the current decreases. The current-voltage characteristics of the sensor is shown in Figure 5.2h. The selectivity test is conducted by exposing sensing region to 500 ppb of gases like carbon monoxide, nitric oxide, and acetone and detected that the sensor is selective to NH_3 only.

The H_2S detection is performed by using PANI/ZnO nanocomposite in the high concentration range ranging from 0.1 to 1,000 ppm [112]. The PANI/ZnO forms a p-n type junction, and hence, it creates difference in their respective Fermi energy. The ZnO is at higher Fermi energy level than PANI, and therefore, easy electron conduction takes place from ZnO to PANI. The observed sensitivity was of the order of 40.5% @ 50 ppm in 63 s with LOD of 100 ppb. Recently, the polymer blend of poly(3-hexylthio- phene-2,5-diyl) (P3HT) and poly(9-vinylcarbazole) (PVK)

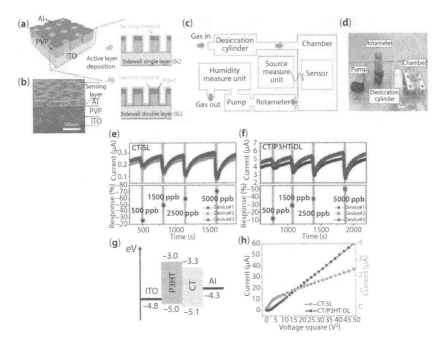

Figure 5.2 (a) Schematic of device structure with sidewall single layer and double layer sensors. (b) SEM cross section image of PBDTTT-C-T DL sensor. (c) The schematic drawing of the measurement system for NH₃ sensing. (d) Picture of measurement system indicating the relative positions of all components. (e) current measurement CT-SL sensors and (f) CT/P3HT-DL sensor exposed to NH3 concentrations of 500 to 5,000 ppb. (g) Energy level diagram of CT/P3HT-DL sensor. (h) Response of current versus voltage square in both CT-SL and CT/P3HT-DL sensors. Reprinted with permission from Ref. [111]. Copyright of American Chemical Society.

OFET-based sensor were successfully employed for NO_2 monitoring by Han *et al.* [113]. The sensor achieved huge NO_2 responsivity of >20,000% for 30 ppm ($\approx$700% for 600 ppb). The response of this blend was 40 times higher than P3HT with achieved LOD was $\approx$300 ppb. Numerous P3HT/PVK blends as 1:0, 1:1, 1:4, and 1:8 were prepared and their sensing response were measured under the exposure of NO_2 in the concentration varied from 0.6 to 30 ppm. It was observed that the doping effect was the main parameter behind this increase in sensitivity.

The H_2 sensing is done by using polypyrrole nanowire of size 40–90 nm in diameter which successfully deposited through template-free electropolymerization [114]. The resistance-based sensor was capable to detect H_2 concentration from 600 to 10,000 ppm. The response of the sensor was found around 3.88 within 72 s. However, sensor possesses very high LOD compared with other resistance-based sensors. The sensor cross-sensitivity was cross-checked under the exposure of CO gas, and it showed very less interference in the H_2 sensitivity. The experimental results are well agreed with the theory proposed for polypyrrole, as under the H_2 environment, the resistance of the PPy is decreased. Another work consists of PMMA/Pd/SLG nanocomposite is fabricated for H_2 sensing [115]. The sensor exhibited response of 66.37% in 1.81 min but observed slightly higher recovery in 5.52 min under the exposure of 2% H_2. Figure 5.3a shows the development of PMMA/Pd/SLG sensor. The sensor is designed by growing graphene via chemical vapor deposition method followed by depositing Pd nanoparticles acts as detecting layer and PMMA as a supporting matrix. The 2% H_2 gas is exposed to sensing region as shown in Figure 5.3b. The Pd has high affinity towards H_2 gas, and after interaction, the volumetric expansion occurs which decreases the resistance and more conductance path is formed, and hence, it further increases the sensitivity of sensor (shown in Figure 5.3c). The time-response of fabricated device is shown in Figures 5.3d and e. The high sensitivity of around "66.38%" is observed with 2% H_2.

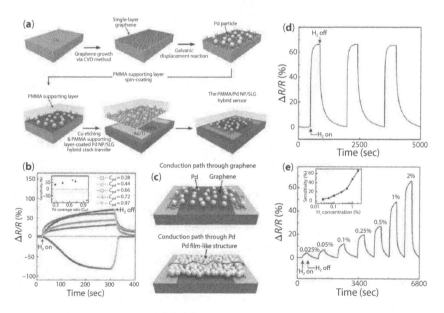

Figure 5.3 (a) Schematic illustration of the procedures used to fabricate the PMMA/Pd NP/SLG hybrid sensor, (b) relative resistance of sensor with different values of capacitance as function of time at 2% H_2, (c) schematic of conduction path through (i) graphene and (ii) Pd, (d) reproducible resistance changes upon exposure to 2% H_2, (e) measurement of percent change in resistance upon exposure to different concentrations of H_2 ranging from 0.025 to 2%. The inset shows a plot of gas response sensitivity with respect to H_2 concentration on a log scale. Reprinted with permission from Ref. [115]. Copyright of American Chemical Society.

We have observed that polymers are extensively studied for various gases detection especially for NH$_3$. Almost all types of nanomaterials including metal oxides, metals, and graphene have been explored for NH$_3$ sensing. It is also pointed out that most of the sensors possess good sensing response but failed in achieving the lower LOD. Certain limitations such as high working temperature of metal oxides and polymer affinity towards humidity need to be considered. In view of the unique features offered by the polymers, very few polymers are exploited for detection of CO, O$_2$, NO$_2$, CH$_4$, etc. Therefore, further study is required on different aspects of polymer-based gas sensing along with the enhancement in responsivity, reliability, and multifunctionality.

5.4.2   Polymer-Based Sensors for Volatile Organic Compounds Detection

Generally, volatile organic compounds (VOCs) are chemical byproducts emitted from industries. The VOCs are extremely hazardous to human health and creates environmental pollution. Some VOCs are carcinogenic and can cause lung cancer. Therefore, their early monitoring becomes very necessary and important for better human health and environmental monitoring. Table 5.2 described the development of some polymer-based VOCs sensing.

Numerous polymers, polymer-nanocomposites, were explored for the detection of various kinds of vapors. A highly luminescent ZnO-polymer nanohybrids were employed for ethanol detection based on photoluminescence [140]. The number of ZnO/PDMS nanocomposite samples of ZnO/PDMS were synthesized and tested for ethanol vapor. The sensor showed LOD around 0.4 Torr (500 ppm). However, the sensor possesses high recovery time of around 7 min. For toluene, most of the polymer composites are demonstrated with increase in sensitivity and response time. Optical Bragg grating principle has been used for toluene finding by utilizing ZnO combined in a polymer DBR structure which consists of alternative layers of polystyrene and cellulose acetate, respectively [141]. The sensor response measured in the range from 20 to 120 ppm of toluene vapor. The sensor showed good resolution below 10 ppm with much better response time of 13 s. Methanol detection is also performed by using polymer like poly(EDOT-co-TAA)]-gold NPs based on chemiresistive principle [142]. The sensor showed sensitivity of 5.09 ± 0.26% at 23,665 ppm of methanol concentration. However, it responded in 130–200 s and recovered in 270–380 s. An infrared spectroscopy has been successfully implemented for the detection of benzene and other type of VOCs

Table 5.2 Polymer-based sensor for volatile organic compound detection.

Materials	Type of sensor	Analyte	Detection range	Sensitivity (S)	Limit of Detection (LOD)	Response/ recovery time	Ref.
Polypyrrole-Polyethylene oxide	Absorbance based	Ethanol	0.1–10 ppm	3.89 @ 10 ppm	–	0.295 s/ 0.418 s	[144]
ZnO–PDMS	Photolumine-scence based	Ethanol	60 to 0.4 Torr [145]		500 ppm	50 s/	[140]
Zinc oxide–polystyrene nanocomposite	Bragg reflector based	Toluene	20–120 ppm		<10 ppm	13 s/-	[141]
PDMS/CNT composite	Resistive based	Toluene	–	0.008 L µg-1	1.5 µgL−1	90 s/30 s	[146]
NH$_2$-MIL-53(Al) MOF mixed in a Matrimid polymer	Capacitive based	Methanol	1,000–2,0000 ppm	–	–	–	[147]
Poly(3,4-ethylenedioxy-thiophene- co -thiophene-3-aceticacid) [poly(EDOT-co-TAA)]-gold NPs	Resistive based	Methanol	23,665 ppm	5.09 ± 0.26	475 ppm	130–200 s/ 270–380 s	[142]

(Continued)

Table 5.2 Polymer-based sensor for volatile organic compound detection. (*Continued*)

Materials	Type of sensor	Analyte	Detection range	Sensitivity (S)	Limit of Detection (LOD)	Response/ recovery time	Ref.
Polydimethylsiloxane (PDMS) with a semimetallic elastomer of poly(3,4-ethylenedioxythiophene):poly(styrenesulfonate) (PEDOT:PSS)	Current based	Acetylene	0–1,000 ppm	89% @1000 ppm	–	-/-	[148]
Polymeric Micelle-Platinum- Decorated Mesoporous TiO$_2$	QCM based	Acetaldehyde	100–500 ppm	4.5 Hz/ppm	27 ppm	-/-	[149]
Polyvinyl ferrocene (PVF)-NaClO$_4$ (PVFoxB)	QCM	Benzene	1%–10%(v/v)	–	<1%	~20 min /~20 min	[150]
(poly)isobutylene (PIB)	Disk Micro-resonators based	Benzene toluene m-xylene	0 ppm to 16,000 ppm	0.0141 Hz/ppm, 0.0414 Hz/ppm, 0.116 Hz/ppm	5.3 ppm, 1.2 ppm, 0.6 ppm	-/-	[151]

(*Continued*)

Table 5.2 Polymer-based sensor for volatile organic compound detection. (*Continued*)

Materials	Type of sensor	Analyte	Detection range	Sensitivity (S)	Limit of Detection (LOD)	Response/ recovery time	Ref.
Poly(N-(1- naphthyl)-N'-(n-octadecyl)carbodiimide (polyNOC)	FTIR Spectroscopy	Benzene vapor	10 mg/ml	Population state= ~ 1.35	–	10 s/-	[143]
Polydiacetylene (PDA)- aerogel	Fluorescence based	Benzene, Acetone, Toluene, 2-propanol	0–1,200 ppm	–	–	1 min /-	[152]
Poly- (vinylpyrrolidone) (PVP)- single-walled carbon nanotubes (SWCNTs)	Resistance based	Isopropyl Alcohol	100–10, 000 ppm	–	–	<3 min /-	[153]
PDMS	Fabry-Perot geometry	Chloroform, Toluene	0–200 ppm	23.7 pm/ppm, 21.6 pm/ppm	840 ppb, 920 ppb	35 s/ 10 s, 57 s/ 25 s	[154]
Poly(HEMA-co- PyMA) polymer	Fluorescence based	2,4,6-trinitrotoluene (TNT)	0–125 ppm	–	–	~5s /-	[155]

(*Continued*)

Table 5.2 Polymer-based sensor for volatile organic compound detection. (*Continued*)

Materials	Type of sensor	Analyte	Detection range	Sensitivity (S)	Limit of Detection (LOD)	Response/ recovery time	Ref.
Polyepichlorohydrin (PECH) and Polyetherurethane (PEUT) with different % of MWCNTs	SAW based	Octane, Toluene	25–200 ppm	1.01 Hz/ppm, 4.38 Hz/ppm	9.2 ppm, 1.7 ppm	30-40 s/ 50-60 s	[57]
P(VBC-co-MMA)-SiO$_2$ IOPC	Reflectance based	o-Xylene, m- Xylene, p- Xylene	–	52 nm	0.51 µg/ml, 0.41 µg/ml, 0.17 µg/ml	–	[156]
Polystyrene nanobeads (PS-NH2)	Absorption based	nitrobenzene (NB)	–	50%	5.6 ppm	30 min /-	[157]
Tris(phenylene)vinylene (TPV)	Fluorescence based	1,3,5-Trinitroperhydro-1,3,5-triazine (RDX) vapor	0–18,765 pg	22 ± 11%	–	30 s /-	[158]
Vertically aligned-carbon nanotubes (VA-CNT)-poly(3,4-ethylenedioxythiophene) (PEDOT)	Resistance based	n-pentane	700–7,000 ppm	0.44	50 ppm	584 s / 374 s	[159]

by using poly(N-(1-naphthyl)-N'-(n-octadecyl)carbodiimide (polyNOC) [143]. Under the exposure of benzene vapor, the molecular structure of polyNOC expands or contracts, and hence, the response in presence of vapor can be measured by using infrared spectroscopy. Under 10 mg/ml of benzene, the population state ratio was reached up to 1.35 with response time of 10 s. Polypyrrole-based sensors have also been explored for vapor sensing [144, 145].

Park *et al.* [160] developed polydiacetylene (PDAs)-based colorimetric sensor for detection of various VOCs by using a smartphone. The sensor is fabricated by designing four PDA arrays on a paper. The PDAs arrays were prepared through photopolymerization as shown in Figure 5.4a. Each consists of alkyl group which is sensitive to specific vapor. When solvent expose to the PDAs array, color of the PDAs array changes simultaneously. As depicted in Figure 5.4b, color of the arrays appeared distinct with respect to each solvent. The UV-absorption is also measured for four PDAs pattern in presence of 11 organic solvents as mentioned above. Figures 5.4c–f show shift in absorption and is occurred due to interaction of four PDSs (1–4) with the different solvents. In this scheme, sensor may give similar colorimetric response over saturation of the vapor phase solvents.

The detection of VOCs is always desirable due to its severe impact on human being and environment. Almost all techniques are widely explored for the detection of vapors. However, very few reports are explained about the specific detection of vapor in the mix environment, and therefore, study on specificity and selectivity needs to be focused.

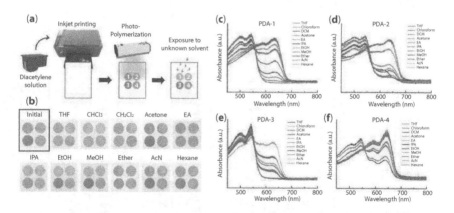

Figure 5.4 (a) Schematic for fabricating and operating the smart sensor system. (b) Different color patterns when the PDA sensors were exposed to various solvents. The UV-vis absorbance spectra for solvatochromic responses of (c) PDA-1, (d) PDA-2, (e) PDA-3, and (f) PDA-4. Reprinted with permission from Ref. [160]. Copyright of American Chemical Society.

5.4.3 Polymer-Based Sensors for Ion Detection

Detection of toxic ions is very significant as it releases to natural sources through many chemical industries in the form of effluent. These ions not only endangered the human life but also aquatic life. Therefore, detection of such ions with fast, sensitive, and reliable sensor is always desirable. Table 5.3 shows the numerous sensors developed for ion sensing in the recent past.

Among the various techniques, optical fluorescence-based technique is widely explored for toxic ion detection. Ding *et al.* [161] demonstrated a fluorescence-based approach for detection of Fe^{3+} ion. A new type of fluorescent conjugated polymer sensing film was fabricated using electropolymerize procedure. The sensor was capable to detect Fe^{3+} ion ranging from 10 µM to 1.3 mM with observed LOD of 5.3 µM. The sensor has shown sensitivity of 20.3 a.u/10^{-5} M. Another polymer film for detection of Hg^{2+} based on colorimetric principle has been demonstrated by Kim *et al.* [162]. The detection is mainly focused on absorption in which the mercury ions interact with the polymer surface. It was observed that the absorption peak was changed from 426 nm to 446 nm in Hg^{2+} in 0.02–0.20 mM. The total wavelength change shift of 20 nm with linear response of wavelength change is observed. The sensor was capable to detect up to 58 µM. The detection of CN^- ion is performed through an optical technique by using N,N-dimethylacrylamide (DMA) and 3-vinylbenzaldehyde (VBA)/(E)-2-amino-4-((4-nitrophenyl)diazenyl)phenol polymer [163]. Due to interaction of CN^- ions with the polymeric structure forms a quinoid structure. As the concentration of CN^- ion varied from 0-28 mM, the corresponding position of absorption wavelength changed from 411 to 530 nm. The sensor shown total shift of 119 nm with observed LOD of 0.1 mM. Thus, the color change from brick-red to purple and can be easily detected through naked eye.

Very recently, PEI-GA-PEI-modified anodic alumina (NAA) interferometer with reflectometric interference spectroscopy (RIfS) has been demonstrated for Cu^{2+} detection [179]. The PEI-GA-PEI shows good response towards Cu^{2+} ion in 1–100 mg/L with LOD around 0.007. Figure 5.5a depicts fabrication of NAA interferometer which is modified with NAA and RIfS. When copper ions interact with sensing area, the optical thickness of film gets altered. The experimental arrangement is shown in Figure 5.5b. Figure 5.5c shows the response of an interferometer with and without Cu^{2+} ion. The effective optical length change with respect to Cu^{2+} ion is depicted in Figure 5.5d. The detailed surface chemistry is explained by the Figures 5.5e and f.

Table 5.3 Polymer-based sensors for various ions detection.

Materials	Type of sensor	Ion type	Detection range	Sensitivity (S)	Limit of Detection (LOD)	Ref.
Silver(I)-based coordination polymer,{[Ag(4-bpmd)]ClO4·DMF}n {[Ag(4-bpmd)] ClO4·DMF}n (1)	Fluorescence based	Fe^{3+}	5–50 μM	–	3.5×10^{-5} mol/L	[164]
{Zn(L)(bpp)+·DMF} n (1) and {[Zn(L) (bpe)+DMF-n (2) (L = 2,2'-[benzene-1,3-diylbis(methanediyl-sulfanediyl)]dibenzoic acid, bpp= 1,3-bis(4-pyridyl)propane, bpe = 1,2-Bis(4-pyridyl) ethylene, DMF = N,N-Dimethylformamide)	Luminescence based	Fe^{3+} and $Cr_2O_7^{2-}$	0–320 μl, 0–125 μl	–	0.76 μM, 3.52 μM	[165]

(*Continued*)

Table 5.3 Polymer-based sensors for various ions detection. (*Continued*)

Materials	Type of sensor	Ion type	Detection range	Sensitivity (S)	Limit of Detection (LOD)	Ref.
9,9′-bis(N-carbazolyl-hexyl)-2-bromofluorene (1), 9,9′-bis(N-carbazolyl-hexyl)-2-ethynylfluorene (3)	Fluorescence based	Fe^{3+}	10 μM to 1.3 mM	20.3 a.u/10^{-5}M	5.3 μM	[161]
Polyethylene terephthalate (PET)-Carbon dots	Photolumine-scence based	Fe^{3+}	0.5–400 μM	–	0.21 μM	[166]
Cd(II)/Zn(II) coordination polymers based on 4,4′-(1H-1,2,4-triazol-1-yl)methylene-bis(benzonic acid)	Luminescence based	Fe^{3+}	0–0.020 mM	–	–	[167]
Poly(azomethine-urethane)	Fluorescence based	Fe^{3+}	1 mM to 0.03125 mM	25,287 a.u./mM	28.90 μM	[168]

(*Continued*)

Table 5.3 Polymer-based sensors for various ions detection. (*Continued*)

Materials	Type of sensor	Ion type	Detection range	Sensitivity (S)	Limit of Detection (LOD)	Ref.
N-[4-(2-Oxo-2H-chromen-3-yl)-thiazol-2-yl]-acrylamide (OCTAA)	Fluorescence based	Hg^{2+}	0.05–1.2 µmol L-1	Imprinting factor of 2.37	0.02 µmol L-1	[169]
Boron dipyrromethene (BODIPY) derivative (9AnPD)	Fluorescence based	Hg^{2+}	1×10^{-4} to 1×10^{-8}	–	1×10^{-6} M	[170]
Terpolymer (P1) of N,N-dimethylacrylamide (DMA), (E)-2-((4-((4-formylphenyl)diazenyl)phenyl)(methyl) amino) ethyl acrylate (FPDEA, M1), and N- (4-benzoylphenyl) acrylamide (BPAm)	Colorimetric based	Hg^{2+}	0.02–0.20 mM	Absorption shift = 20 nm	58 µM	[162]

(*Continued*)

Table 5.3 Polymer-based sensors for various ions detection. (*Continued*)

Materials	Type of sensor	Ion type	Detection range	Sensitivity (S)	Limit of Detection (LOD)	Ref.
N,N-dimethylacrylamide (DMA) and 3-vinylbenzaldehyde (VBA)/(E)-2-amino-4-((4-nitrophenyl) diazenyl)phenol	Colorimetric based	CN^-	0–28 mM	Shift = 119 nm	0.1 mM	[163]
Porphyrin cored polyepichlorohydrin (POR-PECH)	Fluorescence based	CN^-	0–1.2	0.051 µM	–	[171]
Poly{(N-iso-propylacrylamide)-co-(stearyl acid)-co-[9,9-dihexylfluorene-2-bipyridine-7-(4-vinylphenyl)]} (poly (NIPAAm-co-SA-co-FBPY)	Photolumine-scence based	Zn^{2+}	10^{-10}–10^{-3} M	10^{-6} M	–	[172]

(*Continued*)

Table 5.3 Polymer-based sensors for various ions detection. (*Continued*)

Materials	Type of sensor	Ion type	Detection range	Sensitivity (S)	Limit of Detection (LOD)	Ref.
PFQT polymer	Fluorescence based	Zn^{2+}	–	–	10^{-8} mol/L	[69]
Plasticized-PVC/ TiO2-PhC	Fluorescence based	K^+	10^{-5}–10^{-6}	–	–	[173]
Imprinted polymer	Electrochemical impedance spectroscopy (EIS)	nitrate-N	1–10 (mg/L)	–12811 kohm/ (mg/ml)	–	[174]
1,5-naphthyridine-based conjugated polymers	Absorption based	I^-	1–16 ppm	0.0192 a.u./ppm	6.5 ppm	[175]
Fe3O4@SiO2@IIP/ GCE cadmium (II) imprinted polymer nanoparticles (IIP-NPs)	Differential pulse voltammetry based	Cd^{2+}	0.05 to 0.80 µM	–	1×10^{-4} µM	[176]

(*Continued*)

Table 5.3 Polymer-based sensors for various ions detection. (*Continued*)

Materials	Type of sensor	Ion type	Detection range	Sensitivity (S)	Limit of Detection (LOD)	Ref.
Polypyrrole titanium(IV) sulphosalicylo phosphate	Potentiometric based	Pb^{2+}	1×10^{-1} M L^{-1} to 1×10^{-8} M L^{-1}	30 mV decade^{-1}	1×10^{-8} M	[105]
[Cd(bipy)][HL] n (1)(H3L ¼ 20-carboxybiphenyl-4-ylmethylphosphonic acid, and bipy ¼ 2,20-bipyridine)	Fluorescence based	$Cr_2O_7^{2-}$	5×10^{-7} to 5×10^{-4}	$K_{sv} = 9.3 \times 10^{-3}/$ ppm	–	[177]
Carbon nanodots functionalized with rhodamine and poly(ethylene glycol)	Fluorescence based	Al^{3+}	0–10 mM	Intensity ratio: 5×10^{-5} M	1.8×10^{-5} M	[178]

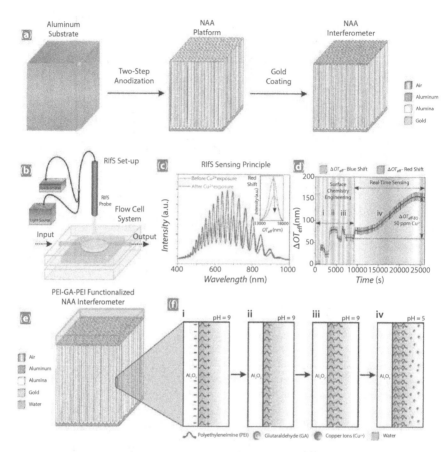

Figure 5.5 (a) Illustration describing the two-step anodization process used to produce NAA interferometers. (b) Schematic showing the RIfS setup used to monitor binding interactions between PEIGA-PEI-modified NAA interferometers and copper ions in real-time under dynamic flow conditions. (c) Spectrum of RIfS of PEI-functionalized NAA interferometers before and after exposure to Cu^{2+} ions. (d) Real-time effective optical thickness changes (ΔOT) associated with the surface chemistry. (e) Schematic showing the structure of PEI-GA-PEI-functionalized NAA interferometers. (f) Illustration showing details of the inner surface chemistry of gold-coated PEI-GA-PEI-functionalized NAA interferometers during different stages of the sensing process (i–iv). Reprinted with permission from Ref. [179]. Copyright of American Chemical Society.

The detection of ions at very low ppm is an important but always a difficult task. The specific immobilization is always essential to increase the sensor performance and selectivity. Most of the polymer-based sensors are focused on detection of Fe^{2+} and Hg^{2+} and very few geometries are explored for detection of other ions. Therefore, it is very crucial to fabricate the sensor which could detect multiple ions with great accuracy and responsivity.

5.4.4 Polymer-Based Sensors for pH Sensing

Monitoring pH level is always a crucial task for human health. Change in pH level can create adverse effects not only on human being but also on ecosystem. Therefore, for better human health and environmental monitoring, pH detection becomes very important aspect.

Various techniques have been implemented for the detection of pH level. Table 5.4 shows some proposed and fabricated polymer-based sensory systems for pH detection. Polymer-based sensors are proved to be very effective due to presence of various groups in their structure. Generally, most common polymers, namely, acidic group: carboxylic acids [−COOH], sulfonic acid [−SO3H], and basic group: nitrogen-containing group is extensively utilized for pH sensing. These kinds of polymers can be obtained from the synthesis procedures or from nature. The acidic or basic group of these polymers gets ionized with surrounding pH change which further alters their structure. However, it is difficult to obtain complete ionization due to an electrostatic effect.

The most common pH sensitive polymer structures with their sensing principle are shown in Figure 5.6. Most of the pH sensors are based on potentiometry, amperometry, voltammetry, and chemoresistance, etc. Many fluorescence-based pH sensors have been explored for biomedical field including drug delivery and magnetic resonance-based applications. A highly sensitive photostable triangulenium fluorophore placed on a polycarbonate substrate was utilized for pH detecting [180]. An optical fiber-based geometry was capable to work in pH ranging from 4.6 to 7.6. The observed responsivity of the polymer was 0.1 optical unit/pH with resolution of 0.03 pH in <90 s was obtained. However, operational range can be adjusted with the incorporation of other dyes. Another luminescence-based pH sensor has been proposed by Duong et al. [181] using FA-HPTS)/3-GPTMS and 3-APTMS sol-gel matrix and a polyurethane hydrogel. The fabricated sensor had detection range from pH 5–8. The developed pH membranes were tested in real wastewater. In all fabricated pH-sensing films, the FA@TEOS-GA = PU and HPTS@K300-GA = PU possesses good sensitivity in pH 5–8. Most recently, Poly-aniline/SU-8 (an epoxy) has been utilized for pH monitoring of red soil based on conductance mechanism [182]. It displays working range in the pH range from 2 to 10 of red soil in just 5–10 s along with measured recovery in 30 s. It possessed response of around ~957 μS/pH in acidic media and for basic medium the sensitivity of ~290 μS/pH was detected. This sensor could be deployed for many agriculture-based applications. The sensing mechanism of Poly-aniline/SU-8 towards varied pH range is occurred due to interaction of amino

Table 5.4 Polymer-coated sensors for pH detection.

Materials	Type of sensor	Detection range	Sensitivity (S)	Limit of Detection (LOD)	Response/ recovery time	Ref.
Diazaoxy-triangulenium (DAOTA) scaffold-siloxane-functionalized linker-Polycarbonate substrate	Fluorescence based	4.6–7.6	~0.1 optical unit per pH unit	0.03 pH	<90 s	[180]
Poly[(9,9-dioctylfluorenyl-2,7-diyl)-co-(1,4-benzo-{2,1′,3}-thiadazole)] (PFBT) Pdots contained tetraphenylporphyrin (TPP)	Fluorescence based	1–13	- ~ 100 nA	–	–	[185]
PEDOT (poly(3,4-ethylenedioxythiophene)) doped with pH dyes (BTB and MO, i.e., Bromothymol Blue and Methyl Orange)	Voltammetric based	1–9	62 ± 2mV pH unit-1, 31 ± 2 mV pH unit -1	–	–	[88]
Lanthanide complex (Eu^{3+}-TTA-PDA)/cellulose acetate (CA)	Luminescence based	2–10	~ 10 a.u./pH	–	–	[186]
Poly-aniline/SU-8 (an epoxy resin) nanocomposite	Conductance based	2.4–10	~957 µS/pH	–	5–10 s/30 s	[182]

(Continued)

Table 5.4 Polymer-coated sensors for pH detection. (*Continued*)

Materials	Type of sensor	Detection range	Sensitivity (S)	Limit of Detection (LOD)	Response/ recovery time	Ref.
Polyaniline (PANI)-polyvinyl alcohol (PVA) optical fiber	Absorption based	2–9	2.79 μW/pH	–	–	[182]
Carbon loaded polystyrene microneedle	Voltammetry based	3.69–7.92	–0.0575	–	–	[187]
Fluorescein amine (FA) and 8-hydroxypyrene-1,3,6-trisulfonic acid trisodium salt (HPTS)/ 3-glycidoxypropyl-trimethoxysilane (GPTMS) and 3-aminopropyltrimethoxysilane (APTMS) sol-gel matrix (GA) and a polyurethane hydrogel (PU)	Fluorescence based	2–6, 5–8	~ 3,500 a.u. @6 pH, ~4,200 a.u. @8 pH	–	–	[181]
Cellulosic paper modified with MMA-SPEA copolymer	Absorption based	1–14	Maximum responsivity ($k_c = 0.14$)	–	–	[188]
PDA films coated on carbon nano-onion modified glassy carbon electrodes (GCE/CNO)	Potentiometric based	1.70–8.36	36 mV/pH	–	–	[189]

(*Continued*)

Table 5.4 Polymer-coated sensors for pH detection. (*Continued*)

Materials	Type of sensor	Detection range	Sensitivity (S)	Limit of Detection (LOD)	Response/recovery time	Ref.
Polyvinyl alcohol (PVA) with polyacrylic acid (PAA)	Thermal actuator based	4–10	−1.25 to −4.4 K/pH (in pH 6-8)	–	–	[190]
Graphite-polyurethane composite	Cyclic voltammetry based	5–9	11.13 ± 5.8 mV/pH	–	8 s/-	[183]
Polyaniline- reduced graphene oxide (ERGO-PA)	Cyclic voltammetry	2–9	55 mV/pH	–	~s/-	[17]
Polyaniline-coated tilted fiber Bragg gratings	Wavelength shift based	2–12	82 pm/pH	–		[13]
Poly(N-(2- hydroxypropyl) methacrylamide) (PHPMA) and integrated/naphthalimide- based/gadolinium− 1,4,7,10-tetraazacyclododecane-1,4,7,10-tetraacetic acid complex (Gd–DOTA complex)	Fluorescence based	3–9	Intensity ratio = ~5	–	–	[191]

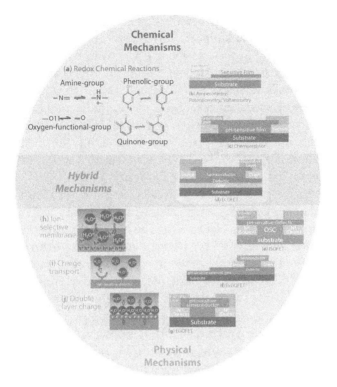

Figure 5.6 A summary of the mechanisms of polymers and organic materials-based pH sensors and their applications in pH sensing devices. (a) Chemical mechanisms for pH sensing, (b) Amperometry/Potentiometry/Voltammetry techniques, and (c) Chemiresistor for chemical mechanisms based pH sensing. (d) ECOFET in hybrid mechanisms-based pH sensing. (e) ISOFET, (f) ExGOFET, and (g) EGOFET in physical mechanism-based pH sensing. (h) Ion-selective membrane, (i) charge transport, and (j) double-layer charge in physical sensing system. Reprinted with permission from Ref. [25]. Copyright of Elsevier.

groups to hydrogen ions present in pH/soil solution. The charge density on PANI composite increases due to interaction of amino group and hence changes in conductance of the sensor increased with increase in pH. Cyclic voltammetry-based graphite-polyurethane composite is explored for sweat pH monitoring [183]. The sensor responded in 11.13 ± 5.8 mV/pH within 8 s. A very negligible intrusion is observed with other types of ions Na$^+$, K$^+$, glucose. The principle of sensing was depending on electro-chemical oxidation reaction between polymer and solution. Optical PANI-TFBG coated pH detection has been demonstrated in pH from 2 to 12 [13]. The response was directly related to film thickness of polymer. The PANI film deposited through oxidative polymerization process. Due to covering of

PANI on cladding of optical fiber, the higher order modes generate due to difference in core and cladding refractive index. This brings the wavelength and intensity change in the interference spectrum which is correlated to the film thickness. It exhibits slope of 82 pm/pH. The temperature effect is also experimentally verified.

SPR sensors are very much sensitive due to interaction of higher order modes with the target molecules through total internal reflection of light which causes formation of evanescent wave along the metal-dielectric boundary. Optical fiber is proved to be very sensitive and versatile method for pH monitoring. Zhao *et al.* [184] fabricated a MMF/SMF/MMF which is layered with hydrogel material with calculated slope of 13 nm/pH during 8–10 pH change. Overall, sensor bandwidth is 1–12. The experimental arrangement is shown in Figure 5.7a. The hydrogel synthesized by using AAM, BAAM, TEMED, and methacrylic acid. The Ag/hydrogel is coated on surface of optical fiber as shown in Figure 5.7b. Due to interaction of hydrogel with either acidic or basic pH, the carboxyl ions get form and therefore significantly swell hydrogel. The swelling causes shift in the resonance peak. Figures 5.7c and d show the response of sensor with and without hydrogel under variation in pH. It is observed that, as the pH value increases, the resonance shows blue shift. Figures 5.7e and f display the

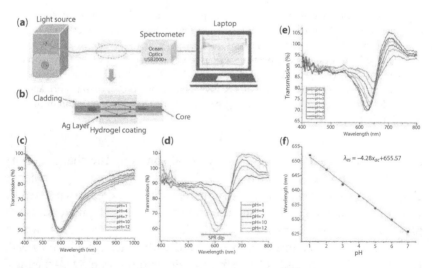

Figure 5.7 (a) Experimental setup of the pH sensing system. (b) Schematic of hydrogel-coating optical fiber SPR sensor. (c) SPR spectra of different pH values without hydrogel. (d) SPR spectra of different pH values with hydrogel. (e) SPR spectra of pH range from 1 to 7 at 22°C (f) Variation of SPR resonance wavelength of pH range from 1 to 7 at 22°C. Reprinted with permission from Ref. [184]. Copyright of Elsevier.

sensor SPR spectra response and SPR resonance wavelength shift in pH range from 1 to 7 at 22°C with obtained response of 4.28 nm/pH. However, proposed sensor possesses certain limitations such as the hydrogel can be sensitive to temperature and hence may affect the sensitivity of the sensor.

Real-time monitoring pH is always required for many different fields like agriculture, biomedical, food technologies, and for environmental monitoring. Number of techniques including chemiresistive and optical-based are widely adopted for pH sensing. Polymers show great affinity towards pH change due to their acidic or basic group. However, while fabricating pH sensor, there is need to be focused on development of selective detection of multiple analytes simultaneously. Various new polymers are projected for biomedical applications but those are needed to be verified *in vitro* and *in vivo*.

5.4.5   Polymer-Based Sensors for Humidity Sensing

Detection of traces of humidity in the environment is a big challenge and creates influence in many fields such as semiconductor industries, chemical industries, aviation, and food packaging industries, etc. Therefore, humidity sensors with insensitive to air pollutants and temperature is always desirable.

Generally, two kinds of polymer such as hydrophilic and hydrophobic are widely utilized for humidity sensing. It is detected that the hydrophilic polymer owns small diffusion coefficients because of strong H-bond interaction than hydrophobic polymer. Therefore, the hydrophilic polymer shows high response than the hydrophobic polymer. In addition to this, due to advancement of nanomaterials, several properties of polymer-nanocomposite could be tailored. Nanoparticles possess large surface area and therefore could detect fine traces of humidity with great accuracy. Many approaches have been explored towards the detection of humidity with wide range of RHs as shown in Table 5.5. The hydrophobic PVDF polymer converted into the hydrophilic nature, and it was employed for humidity sensing [192]. The hydrophilic nature TiO_2-PVDF has shown increased in sensitivity with linear response to humidity in 20%–95% RH. This capacitive-based sensor shows the sensitivity of ~1pF in the humidity range from 30%–80%. The addition of TiO_2 NPs increased the hydrophilicity of the polymer nanocomposite. The experimental results confirmed that the PVDF-TiO2 of concentration in "2.5 wt%–wt 0.5%" shown increased in sensor response than that of PVDF-TiO2 in concentration varied from "5 wt%–wt 0.5%". The sensor exhibits the response of 45 s and fast recovery of 11 s, respectively. The PEDOT:PSS resistance-based sensor has been proposed and demonstrated for humidity sensing in wide range from 10% to

Table 5.5 Polymer-based sensors for humidity sensing.

Materials	Type of sensor	Detection range (in RH)	Sensitivity (S)	Limit of Detection (LOD)	Response time/ recovery time	Ref.
Polyvinylidene fluoride titanium dioxide (PVDF-TiO$_2$) nanocomposite	Capacitive based	30%–80%	~1 pF	–	45 s/11 s	[192]
Humidity Sensitive Polymers (HSPs)/ Humidity Insensitive Polymer (HIP)	Refractive index change based	20%–100%	RI Change = 0.0016	–	–	[196]
PEDOT:PSS	Resistance based	10%–90%	3.7125 kΩ/RH	–	/30 s	[193]
Sulfonated poly(ether ether ketone) (SPEEK)-metal–organic–framework	Impedance based	11%–95%	5.46	–	9 s/130 s	[197]
Poly(methyl methacrylate) (PMMA) and polyaniline (PANI)	Absorption based	33%–98%	Change in extinction = ~0.04	–	1–2 min/-	[194]

(Continued)

Table 5.5 Polymer-based sensors for humidity sensing. (*Continued*)

Materials	Type of sensor	Detection range (in RH)	Sensitivity (S)	Limit of Detection (LOD)	Response time/ recovery time	Ref.
Cadmium Selenide Quantum Dots/ Poly-(dioctylfluorene)	Capacitive based	10%–90%	3.30	–	9 s/7 s	[21]
PEDOT: PSS	Resistance based	10%–50%	6.43%	–	–	[198]
Polypyrrole/Tantalum pentoxide (PTO) composite	Impedance based	10%–95%	0.044 Ω/% RH	4.22%RH	6s/7s	[195]
Zirconium phosphonates/ polyaniline	Resistance based	10%–98%	log R= 11 Ω	–	–	[199]
Poly(3,4-ethylenedioxy- thiophene)– poly(styrenesulfonate) (PEDOT:PSS)	Impedance based	0%–100%	~45%	–	–	[92]

(*Continued*)

Table 5.5 Polymer-based sensors for humidity sensing. (*Continued*)

Materials	Type of sensor	Detection range (in RH)	Sensitivity (S)	Limit of Detection (LOD)	Response time/recovery time	Ref.
ZnCo2O4/polypyrrole nanofil	QCM based	0%–100%	58.4 Hz/%RH	–	8 s/7 s	[200]
PS/DCM microspheres	whispering-gallery mode (WGM)	0%–97%	6 pm/RH%	1.4 RH%	–	[201]
Polyethyleneglycol-diacrylate (PEG-DA)	quartz tuning forks (QTFs)	~28 to ~55 %	2.8 Hz/RH%	0.36 RH%	–	[202]
Polyaniline (PANI) and-carbon nanofiber (CNF)- polyvinyl alcohol (PVA)	Capacitance based	30%–100%	3400 pF	–	41 s/46 s	[203]
Zinc oxide and polypyrrole composite (ZnO/PPy)	Resistance based	5%–95%	0.31/RH	–	12 s/8s	[102]

(*Continued*)

Table 5.5 Polymer-based sensors for humidity sensing. (*Continued*)

Materials	Type of sensor	Detection range (in RH)	Sensitivity (S)	Limit of Detection (LOD)	Response time/ recovery time	Ref.
LiCl-loaded poly(3-Hydroxybenzoic acid) (P3HBA)	Impedance based	11%–95% RH.	16,000 K Ω	–	2s/7s	[204]
Sulfonated Poly (ether ether ketone) (SPEEK)	Resistance based	30%–40% and 60%–90%	−0.0014 MΩ/RH	–	20 s/-	[205]
Polyacrylamide (PAM) microfibers	Wavelength shift based	5%–71%	490pm=%RH		120 ms/-	[206]
Carbon nanotube (ACNT) decoration onto the thermoplastic polyurethane (PU) nanofiber	Resistance based	11%–95%	29%	–	–	[207]

(*Continued*)

Table 5.5 Polymer-based sensors for humidity sensing. (*Continued*)

Materials	Type of sensor	Detection range (in RH)	Sensitivity (S)	Limit of Detection (LOD)	Response time/ recovery time	Ref.
Carbon nanocoils (CNCs)- liquid crystal polymer (LCP) substrate	Resistance based	4%–80%	0.15%	<4%RH	1.9 s/1.5 s	[208]
Gold nanoparticles (GNPs)/α- Methoxypoly (ethylene glycol)-ω-(11-mercaptoundecanoate) (PEG)	Resistance based	1.8 to 95 RH%	Total change = 10^5	1.8 RH%	≤1.2 s/≤3 s	[209]
Poly(acrylamide-N,N′-methylene bis(acrylamide)) (P(AM-MBA)) nanogels and TiO$_2$ nanoparticles are	Colorimetric based	47.0% to 89.3%	Wavelength change = 426 nm to 668 nm	–	0.1 s/-	[210]

90% [193]. It was observed that when humidity level increased from 10% to 90%, the corresponding resistance also increased. The sensor possesses sensitivity of 3.7125 kΩ/RH. It is observed that PANI and PMMA mostly studied for humidity measurements due to their well stability, conducting nature, and simple synthesis method. The fabrication of PANI-PMMA was done through electrospinning by depositing on the glass and utilized for humidity sensing in the 33%–98% RH level [194]. The sensitivity of the sensor was measured from change in optical extinction coefficient in varied RH level. The results showed that with increasing UV exposure, the extinction coefficient got decreased. During 33%–98% humidity level, the 2.0 wt% PANI-PMMA exhibits ~0.04 change in the extinction within 1–2 min. The impedance-based technique has been explored for wide RH level detection. The polypyrrole has great degree of swelling property in presence of water. The Polypyrrole/Tantalum pentoxide (PTO) composite has shown great affinity towards humidity [195]. The sensor showed sensitivity of 0.044 Ω/% RH with LOD of 4.22% RH in the humidity level of 10%–95%. The PPy is a p-type semiconductor and Ta_2O_5 is a n-type semiconductor which form an interfacial p-n hetero-junction. Therefore, due to formation of covalent bonds in this structure, the chances of water adsorption increase. The metal organic framework derivative $ZnCo_2O_4$/ PPy-based QCM sensor has been proposed for humidity monitoring by Zhang et al. [195]. The QCM-based sensor exhibited good response compared with both pure PPy and pure $ZnCo_2O_4$ sensor in the humidity level from 0% to 100%. The sensor displayed higher sensitivity of 58.4 Hz/%RH in 8 s and recovered in 7 s.

Optical fiber-based sensors also exhibited good response towards humidity. Due to adsorption humidity on sensing material, the material refractive index modifies and will modulate the dip in an interference pattern. Recently, the polymer microcavity fiber Fizeau interferometer (PMFFI) consisting a FBG-based humidity measurement has been carried out [211]. The measurement set up for humidity is shown in Figure 5.8a. The inset of Figure 5.8a shows that the PMFFI distal end is coated with NOA polymer by implementing UV treatment. The measurement is carried out in the humidity level ranging from 20% to 90% (Figure 5.7b). The inset of Figure 5.7b shows the shift in the FBGs spectrum with increase in humidity level. The good linear response of wavelength shift Vs RH is shown in Figure 5.7c in adsorption and desorption mode. The PMFFI sensor shows the sensitivity 0.0545 nm/%RH. Only FBGs response is also measured in presence of humidity but observed to be insensitive to humidity variation. The sensing mechanism can be explained based on swelling of NOA. Due to adsorption of humidity on to the surface of NOA polymer

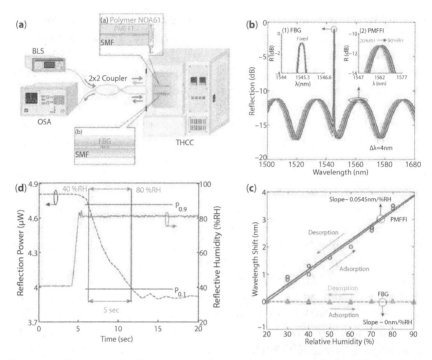

Figure 5.8 (a) Experimental setup for simultaneously measuring RH. (b) Reflection spectra of L = 27 μm PMFFI that incorporates FBG as RH increases. Insets (1) and (2) display shifts of peak wavelengths of FBG and PMFFI. (c) Sensitivities of proposed PMFFI and FBG to RH during adsorption and desorption processes. (d) Response transient of the sensor (dashed line) to increases in RH from 40% to 80% (solid line) at a fixed T of 25°C. Reprinted with permission from Ref. [211]. Copyright of Elsevier.

coating, the surface of NOA swells or shrinks and thus alters the Fabry-Perot cavity length and phase difference created inside the cavity. The measured response time is around 5 s to 40% RH difference as shown in Figure 5.7d.

The polymer-sensors exhibit high responsivity, small hysteresis, better stability, and very fast response time. The recovery of the sensor is quite fast as observed in seconds. Various new techniques and surface morphology plays an important role in humidity sensing. Various polymers/nanomaterials have attracted great attention towards humidity sensing. However, there are certain problems such as most of the polymers possess both water absorption coefficient and thermal expansion coefficient. Therefore, the cross-sensitivity issue may arise and could degrade the sensor performance. Thus, there is tremendous need to develop a highly sensitive and reliable humidity sensor that can operate in harsh environment.

5.5   Outlook and Perspectives

This chapter reviewed polymer-based chemical sensors for many parameters sensing including gases, vapors, ions, pH, and humidity for numerous applications in the field of healthcare, food storage, sewage water treatment, and environmental monitoring. The polymer-based technologies have many advantages such as lightweight, simple, flexible, and sensitive to analyte. Furthermore, the polymers can be tunable, and hence, mixing ratio can be modified accordingly. Thoroughly, we discussed about the advantages and disadvantages of polymers for gases, vapors, ions, and humidity sensing. Numerous immobilization techniques are presented for polymer-based chemical sensor development. Polymers-based sensors are proved to be very effective for chemical sensing. However, there are still certain limitations which need to be improved for better sensory performance.

Polymers are successfully utilized for gas and vapor sensing and displayed remarkable sensitivity and durability. However, polymer-based gas and vapor sensors faced certain limitations such as selectivity and LOD. Most of the sensor is not reached up to ppb or sub-ppb level and only few reported the ppb limit. In most of the polymer-based chemical sensors, the selectivity is still a challenge. Detection of specific analyte in complex environment is always required. This could be the challenging task, and therefore, proper functionalization techniques need to be implemented properly. The binding of specific analyte for the detection of specific target molecule could increase the selectivity and lower the LOD of sensor. Therefore, proper immobilization of appropriate receptors on the surface of sensing region could increase the affinity of polymer towards the target analyte and could increase the selectivity of the sensor. To resolve this issue, the LBL and affinity interactions methods could be useful.

In case of gas and vapor sensors, the sensing mechanism could be well explained by using diffusion and HSP mechanism. When any gas or vapor interacts with the polymer sensing film, it diffuses inside the film and changes the properties of sensing polymer material. It is observed that most of the polymer-based gas or vapor sensors showed high response time, and therefore, one should consider the parameters such as sensing film thickness along with the diffusion coefficient. Therefore, sensor with higher thickness film shows slow response rate than the lower thickness film. Very few have considered this sensing mechanism and hence need to be studied.

It is observed that most of the sensors are based on CPs and are of growing interest for researcher's due to versatility, molecular designs, and signal amplification effect. Some of the methods like electrostatic assembly, electrospinning, and spin-coating are explored for fabrication of CPs. However, it has certain drawbacks such as achieving good morphology control and required large quantity of material. However, most CPs possess less selectivity towards specific analyte. Therefore, need to functionalize with appropriate receptors. The drawback of conducting polymers is that it may suffer degradation problem even in dry atmosphere. Therefore, versatile approach, combinatorial synthesis, and optimization of sensing parameters need to be explored.

It is observed that metal oxides are very sensitive to gases due to their high conductivity and are highly efficient at higher working temperature but fail at low working temperature. The polymer-nanomaterial composite system has great potential to solve the issue of conductivity by incorporating certain metals, CNTs or graphene into it. Therefore, polymer has ability to form a matrix very easily which is sustainable, durable, and more efficient over the time. However, the working environment (humidity, temperature, atmospheric pressure) may affect the sensor performance. With the advancements of nanomaterials and nanotechnology, great efforts have been taken to increase the sensor performance in terms of response/ recovery time, sensitivity, selectivity, and long-term stability. Though, a more sensitive, reliable, low power, and miniaturized sensory system is always desirable.

For polymer-based chemical sensor, numerous methods based on electrical and optical are widely explored. Various methods including resistive based, potentiometric, voltammetry, impedance-based, and amperometry and optical methods including light-absorption, interferometric-based, SPR, and colorimetric methods/techniques have been studied and implemented successfully. Every method has certain merits and demerits, and therefore, proper utilization of techniques along with understanding of surface chemistry of analyte with the sensing material is required to be analyzed.

Polymer-based sensor technologies have many advantages such as its properties can be tailored very easily, simple fabrication techniques, and technologies are cost-effective. Therefore, significant attention need to be focused towards the development of new material, effective optimization of sensor parameters, miniaturized designs, low power consumption, durable approach for chemical sensor, and in commercialization of these technologies.

References

1. Persaud, K.C., Polymers for chemical sensing. *Mater. Today*, 8, 38, 2005.
2. Mcquade, D., Taylor, D., Pullen, A.E., Swager, T.M., Conjugated Polymer-Based Chemical Sensors. *Chem. Rev.*, 100, 2537, 2000.
3. Das, R., Pattanayak, A.J., Swain, S.K., Polymer nanocomposites for sensor devices, pp. 205–218, *Woodhead Publishing*, United Kingdom, Elsevier Ltd., 2018.
4. Islam, M.R., Lu, Z., Li, X., Sarker, A.K., Hu, L., Choi, P., Li, X., Hakobyan, N., Serepe, M.J., Responsive Polymers for Analytical Applications: A Review. *Anal. Chem. Acta*, 789, 17, 2013.
5. Lange, U., Roznyatovskaya, N.V., Mirsky, V.M., Conducting polymers in chemical sensors and arrays. *Anal. Chem. Acta*, 614, 1, 2008.
6. Reyes-Ortega, F., pH-Responsive polymers: Properties, synthesis, and applications, pp. 45–92, Second Edition, *Woodhead Publishing*, United Kingdom, Elsevier Ltd., 2019.
7. Bari, S.S., Chatterjee, A., Mishra, S., Biodegradable polymer nanocomposites: An overview. *Polym. Rev.*, 56, 287, 2016.
8. Mohr, G.J., Polymers for optical sensors, Optical Chemical Sensors. *NATO Science Series II*, 224, 297–321, 2006.
9. Cichosz, S., Masek, A., Zaborski, M., Polymer-based sensors: A review. *Polym. Test.*, 67, 342, 2018.
10. Pavase, T.R., Lin, H., Shaikh, Q., Hussain, S., Li, Z., Ahmed, I., Lv, L., Sun, L., Shah, S., Lalhoro, M.T., Recent advances of conjugated polymer (CP) nanocomposite-based chemical sensors and their applications in food spoilage detection: A comprehensive review. *Sens. Actuators B*, 273, 1113, 2018.
11. Wu, W., Bazan, G., Liu, B., Conjugated-polymer-amplified sensing, imaging, and therapy. *Chem Cell Press*, 2, 760, 2017.
12. Mishra, S.K., Tripathi, S.N., Choudhary, V., Gupta, B.D., SPR based fibre optic ammonia gas sensor utilizing nanocomposite film of PMMA/reduced graphene oxide prepared by in situ polymerization. *Sens. Actuators B*, 199, 190, 2014.
13. Aldaba, A.L., González-Vilab, A., Debliquyc, Lopez-Amo, M., Caucheteur, C., Lahem, D., Polyaniline-coated tilted fiber Bragg gratings for pH sensing. *Sens. Actuators B Chem.*, 254, 1087, 2017.
14. Yeo, T.L., Sun, T., Grattan, K.T.V., Parry, D., Lade, R., Powell, B.D., Characterisation of a polymer-coated fibre Bragg grating sensor for relative humidity sensing. *Sens. Actuators B Chem.*, 110, 148, 2005.

15. Chen, S. and Sun, G., High sensitivity ammonia sensor using a hierarchical polyaniline/poly(ethylene-co-glycidyl methacrylate) nanofi brous composite membrane. *ACS Appl. Mater. Interfaces*, 514, 6473, 2013.
16. Manjunatha, S., Machappa, T., Ravikiran, Y.T., Chethan, B., Sunilkumar, A., Polyaniline based stable humidity sensor operable at room temperature. *Physica B Condens. Matter*, 561, 170, 2019.
17. Chinnathambi, S. and Euverink, G.J.W., Polyaniline functionalized electro-chemically reduced graphene oxide chemiresistive sensor to monitor the pH in real time during microbial fermentations. *Sens. Actuators B Chem.*, 264, 38, 2018.
18. Gross, P., Jaramillo, T., Pruitt, B., Cyclic-Voltammetry-Based Solid-State Gas Sensor for Methane and Other VOC Detection. *Anal. Chem.*, 90, 6102, 2018.
19. Liu, S.G., Liu, T., Li, N., Geng, S., Lei, J.L., Li, N.B., Luo, H.Q., Polyethylenimine-derived fluorescent nonconjugated polymer dots with reversible dual-signal ph response and logic gate operation. *J. Phys. Chem. C*, 121, 6874, 2017.
20. Li, X.S., An, J., Zhang, H., Liu, J., Li, Y., Du, G., Wu, X., Fei, L., Lacoste, J., Cai, Z., Liu, Y., Huo, J., Ding, B., Cluster-based CaII, MgII and CdII coordina-tion polymers based on amino-functionalized tri-phenyl tetra-carboxylate: Bi-functional photo-luminescent sensing for Fe3+ and antibiotics. *Dyes Pigm.*, 170, 107631, 2019.
21. Muhammad, F., Tahir, M., Zeb, M., Wahab, F., Kalasad, M.N., Khan, D.N., Karimov, K.S., Cadmium selenide quantum dots: Synthesis, characterization and their humidity and temperature sensing properties with poly-(dioctyl-fluorene). *Sens. Actuators B Chem.*, 285, 504–512, 2019.
22. Ohira, S., Goto, K., Toda, K., Dasgupta, P.K., A Capacitance sensor for water: Trace moisture measurement in gases and organic solvents. *Anal. Chem.*, 84, 8891, 2012.
23. Sanjuán, A.M., Ruiz, J.A.R., García, F.C., García, J.M., Recent developments in sensing devices based on polymeric systems. *React. Funct. Polym.*, 133, 103–25, 2018.
24. Cinti, S., Polymeric materials for printed-based electroanalytical (bio) appli-cations. *Chemosensors*, 5, 31, 2017.
25. Ul, A., Qin, Y., Nambiar, S., Yeow, J.T.W., Howlader, M.M.R., Hu, N. *et al.*, Polymers and organic materials-based pH sensors for healthcare applica-tions. *Prog. Mater. Sci.*, 96, 174–216, 2018.
26. Westmacott, K., Weng, B., Wallace, G.G., Killard, A.J., Nanostructured con-ducting Polymers for electrochemical Sensing and biosensing, *Woodhead Publishing Limited*. United Kingdom, 150–194, 2014.
27. Carvalho, W.S.P., Wei, M., Ikpo, N., Gao, Y., Serpe, M.J., Polymer-based tech-nologies for sensing applications. *Anal. Chem.*, 90, 459, 2018.
28. Wang, X. and Wolfbeis, O.S., Fiber-Optic chemical sensors and biosensors (2008–2012). *Anal. Chem.*, 85, 487–508, 2013.
29. Ruiz, J., Vallejos, S., Garcia, J., Garcia, J.M., Polymer-Based Chemical Sensors, *Chemosensors*, 6, 42, 2018.

30. Alvarez, A., Costa-Fernández, J., Pereiro, R., Sanz-medel, A., Salinas-castillo, A., Fluorescent conjugated polymers For chemical and Biochemical sensing. *TrAC Trends in Analytical Chemistry*, 30, 1513–25, 2011.

31. Wang, X. and Wolfbeis, O.S., Fiber-Optic Chemical Sensors and Biosensors (2013–2015). *Anal. Chem.* 88, 203–227, 2016.

32. Rivero, P., Goicoechea, J., Arregui, F., Optical fiber sensors based on polymeric sensitive coatings. *Polymers*, 1, 26, 2018.

33. Ruiz, J., Sanjuan, J., Vallejos, S., Garcia, F., Garcia, J.M., Smart Polymers in Micro and Nano Sensory Devices, *Chemosensors*, 6, 12, 2018.

34. García, F.C. and García, J.M., Smart Polymers for Highly Sensitive Sensors and Devices: Alternatives, PP. 607–650, *Woodhead Publishing*, United Kingdom, Second Edi, Elsevier Ltd., 2019.

35. Hu, J. and Liu, S., Responsive polymers for detection and sensing applications: Current status and future developments. *Macromolecules*, 43, 8315, 2010.

36. Canfarotta, F., Whitcombe, M.J., Piletsky, S.A., Polymeric nanoparticles for optical sensing. *Biotechnol. Adv.*, 2013.

37. García, J.M., Pablos, J.L., García, F.C., Serna, F., Sensory Polymers for Detecting Explosives and Chemical Warfare Agents. *Industrial Applications for Intelligent Polymers and Coatings*, 553–76, 2016.

38. Yin, M., Gu, B., An, Q., Yang, C., Liang, Y., Yong, K., Recent development of fiber-optic chemical sensors and biosensors: Mechanisms, materials, micro/nano-fabrications and applications. *Coord. Chem. Rev.*, 376, 348–92, 2018.

39. Bai, H. and Shi, G., Gas Sensors Based on Conducting Polymers. *Sensors*, 7, 267–307, 2007.

40. Yang, Y., Li, S., Yang, W., Yuan, W., Xu, J., Jiang, Y., *In Situ* Polymerization Deposition of Porous Conducting Polymer on Reduced Graphene Oxide for Gas Sensor. *ACS Appl. Mater. Interfaces*, 6, 13807–13814, 2014.

41. Carne, A., Carbonell, C., Imaz, I., Maspoch, D., Nanoscale metal – organic materials. *Chem. Soc. Rev.*, 40, 291–305, 2011.

42. Khan, R.R., Kang, B., Lee, S., Kim, S., Yeom, H., Lee, S. *et al.*, Fiber-optic multi-sensor array for detection of low concentration volatile organic compounds. 21, 8643–53, 2013.

43. Liu, B., Metal–organic framework-based devices: Separation and sensors. *J. Mater. Chem.*, 22, 10094–10101, 2012.

44. Richardson, J.J., Björnmalm, M., Caruso, F., Technology-driven layer-by-layer assembly of nanofilms. *Science (80-)*, 348, 2491-1–11, 2015.

45. Zhang, D., Jiang, C., Zhou, Q., Layer-by-layer self-assembly of tricobalt tetroxide-polymer nanocomposite toward high-performance humidity-sensing. *J. Alloys Compd.*, 711, 652–8, 2017.

46. Cui, S., Yang, L., Wang, J., Wang, X., Fabrication of a sensitive gas sensor based on PPy/TiO 2 nanocomposites films by layer-by-layer self-assembly and its application in food storage. *Sens. Actuators B Chem.*, 233, 337–46, 2016.

47. Tian, F., Kanka, J., Sukhishvili, S.A., Du, H., Photonic crystal fiber for layer-by-layer assembly and measurements of polyelectrolyte thin films. *Opt. Lett.*, 37, 4299–301, 2012.

48. Ji, Q., Honma, I., Paek, S.M., Akada, M., Hill, J.P., Vinu, A. *et al.*, Layer-by-layer films of graphene and ionic liquids for highly selective gas sensing. *Angew. Chem. – Int. Ed.*, 49, 9737–9, 2010.

49. Goicoechea, J., Zamarreño, C.R., Matías, I.R., Arregui, F.J., Optical fiber pH sensors based on layer-by-layer electrostatic self-assembled Neutral Red. *Sens. Actuators B Chem.*, 132, 305–11, 2008.

50. Lobnik, A., Oehme, I., Murkovic, I., Wolfbeis, O.S., pH optical sensors based on sol-gels: Chemical doping versus covalent immobilization. *Anal. Chim. Acta*, 367, 159–65, 1998.

51. Charych, D.H., Nagy, J.O., Spevak, W., Ager, J., Bednarski, M.D., Direct colorimetric detection of virus by a polymerized bilayer assembly. *Mater. Res. Soc. Symp. – Proc.*, 330, 295–307, 1994.

52. Wang, P., Liu, M., Kan, J., Amperometric phenol biosensor based on polyaniline. *Sens. Actuators B Chem.*, 140, 577–84, 2009, doi: 10.1016/j.snb.2009.05.005.

53. Ramanathan, K., Bangar, M.A., Yun, M., Chen, W., Myung, N.V., Mulchandani, A., Bioaffinity sensing using biologically functionalized conducting-polymer nanowire. *J. Am. Chem. Soc.*, 127, 496–7, 2005.

54. Saberi, R.S., Shahrokhian, S., Marrazza, G., Amplified electrochemical DNA sensor based on polyaniline film and gold nanoparticles. *Electroanalysis*, 25, 1373–80, 2013, doi: 10.1002/elan.201200434.

55. Wang, L., Hua, E., Liang, M., Ma, C., Liu, Z., Sheng, S. *et al.*, Graphene sheets, polyaniline and AuNPs based DNA sensor for electrochemical determination of BCR/ABL fusion gene with functional hairpin probe. *Biosens. Bioelectron.*, 51, 201–7, 2014.

56. Wang, T., Green, R., Guldiken, R., Mohapatra, S., Mohapatra, S., Multiple-layer guided surface acoustic wave (SAW)-based pH sensing in longitudinal FiSS-tumoroid cultures. *Biosens. Bioelectron.*, 124–125, 244–52, 2019.

57. Sayago, I., Fernández, M.J., Fontecha, J.L., Horrillo, M.C., Vera, C., Obieta, I. *et al.*, New sensitive layers for surface acoustic wave gas sensors based on polymer and carbon nanotube composites. *Sens. Actuators B Chem.*, 175, 67–72, 2012.

58. Buvailo, A.I., Xing, Y., Hines, J., Dollahon, N., Borguet, E., TiO2/LiCl-Based Nanostructured Thin Film for Humidity Sensor Applications, *ACS Appl. Mater. Interfaces*, 528–33, 2011.

59. Toniolo, R., Pizzariello, A., Dossi, N., Lorenzon, S., Abollino, O., Bontempelli, G., Room Temperature Ionic Liquids As Useful Overlayers for Estimating Food Quality from Their Odor Analysis by Quartz Crystal Microbalance Measurements, *Anal. Chem.* 85, 7241, 2013.

60. Apodaca, D.C., Pernites, R.B., Ponnapati, R.R., Mundo, F.R. Del, Advincula, R.C., Electropolymerized Molecularly Imprinted Polymer Films of a

Bis-Terthiophene Dendron: Folic Acid Quartz Crystal Microbalance Sensing, *ACS Appl. Mater. Interfaces*, 3, 191–203, 2011.

61. Lee, S., Takahara, N., Korposh, S., Yang, D., Toko, K., Kunitake, T., Nanoassembled Thin Film Gas Sensors. III. Sensitive Detection of Amine Odors Using TiO 2/Poly (acrylic acid) Ultrathin Film Quartz Crystal Microbalance Sensors. *Anal. Chem.*, 82, 2228–36, 2010.

62. Wang, S., Liu, J., Zhao, H., Guo, Z., Xing, H., Gao, Y., Electrically Conductive Coordination Polymer for Highly Selective Chemiresistive Sensing of Volatile Amines, *Inorg. Chem.* 57, 541–544, 2018.

63. Takahashi, T., Inoue, A., Yan, H., Kanai, M. *et al.*, Paper-Based Disposable Molecular Sensor Constructed from Oxide Nanowires, Cellulose Nanofibers, and Pencil-Drawn Electrodes. *ACS Appl. Mater. Interfaces*, 11, 15044–50, 2019.

64. Wu, J., Wu, Z., Han, S., Yang, B., Gui, X., Tao, K. *et al.*, Extremely Deformable, Transparent, and High-Performance Gas Sensor Based on Ionic Conductive Hydrogel. *ACS Appl. Mater. Interfaces*, 11, 2364–73, 2018.

65. Li, R.W.C., Ventura, L., Gruber, J., Kawano, Y., Carvalho, L.R.F., A selective conductive polymer-based sensor for volatile halogenated organic compounds (VHOC). 131, 646–51, 2008.

66. Xu, X. and Yan, B., Eu(III)-Functionalized MIL-124 as Fluorescent Probe for Highly Selectively Sensing Ions and Organic Small Molecules Especially for Fe(III) and Fe(II), *ACS Appl. Mater. Interfaces* 7, 721–729, 2015.

67. Zhang, L., Li, T., Li, B., Li, J., Wang, E., Carbon nanotube – DNA hybrid fluorescent sensor for sensitive and selective detection of mercury (II) ion, *Chem. Commun.*, 46, 1476–8, 2010.

68. Huang, X., Meng, J., Dong, Y., Cheng, Y., Zhu, C., Polymer-based fluorescence sensor incorporating triazole moieties for Hg 2 þ detection via click reaction. *Polymer (Guildf)*, 51, 3064–7, 2010.

69. Diao, H., Guo, L., Liu, W., Feng, L., A Novel Polymer Probe for Zn(II) Detection with Ratiometric Fluorescence Signal. *Spectrochim. Acta Part A Mol. Biomol. Spectrosc.*, 196:274–280, 2018.

70. Bogale, R.F., Chen, Y., Ye, J., Yang, Y., Rauf, A., Duan, L. *et al.*, Highly selective and sensitive detection of 4-nitrophenol and Fe3+ ion based on a luminescent layered terbium (III) coordination polymer, *Sens. Act. B.*, 245, 171, 2017.

71. Mishra, S.K. and Gupta, B.D., Surface plasmon resonance based fiber optic sensor for the detection of CrO42-using Ag/ITO/hydrogel layers. *Anal. Methods*, 6, 5191–7, 2014.

72. Song, X., Zhang, M., Wang, C., Shamshooma, A., Meng, H., Xi, W., Mixed lanthanide coordination polymers for temperature sensing and enhanced NdIII NIR luminescence. *J. Lumin.*, 201, 410–418, 2018.

73. Yuan, W., Khan, L., Webb, D.J., Kalli, K., Rasmussen, H.K., Stefani, A. *et al.*, Humidity insensitive TOPAS polymer fiber Bragg grating sensor. 19, 660–2, 2011.

74. Jayabal, S., Sathiyamurthi, R., Ramaraj, R., Selective sensing of Hg2+ ions by optical and colorimetric methods using gold nanorods embedded in a functionalized silicate sol–gel matrix, *J. Mater. Chem. A*, 2, 8918–25, 2014.

75. Yapor, J.P., Alharby, A., Gentry-Weeks, C., Reynolds, M.M., Mashud Alam, A.K.M., Y.V.L., Polydiacetylene Nano fiber Composites as a Colorimetric Sensor Responding To Escherichia coli and pH. *ACS Omega*, 2, 7334–42, 2017.

76. Ullman, A.M., Jones, C.G., Doty, F.P., Stavila, V., Talin, A.A., Allendorf, M.D., Hybrid Polymer/Metal–Organic Framework Films for Colorimetric Water Sensing over a Wide Concentration Range. *ACS Appl. Mater. Interfaces*, 10, 24201–8, 2018.

77. Hui, S., Yap, K., Chien, Y., Tan, R., Rahman, A., Bin, J.W. *et al.*, An Advanced Hand-Held Microfiber-Based Sensor for Ultrasensitive Lead Ion Detection. *ACS Sens.*, 3, 2506–12, 2018.

78. Lee, J.S., Jun, J., Shin, D.H., Jang, J., Urchin-like polypyrrole nanoparticles for highly sensitive and selective chemiresistive sensor application. *Nanoscale*, 6, 4188–94, 2014.

79. Singhal, A., Kaur, M., Dubey, K.A., Bhardwaj, Y.K., Jain, D., Pillai, C.G.S. *et al.*, Polyvinyl alcohol – In2O3 nanocomposite films: Synthesis, characterization and gas sensing properties. *RSC Adv.*, 2, 7180–9, 2012.

80. Do, J. and Wang, S., On the Sensitivity of Conductimetric Acetone Gas Sensor Based on Polypyrrole and Polyaniline Conducting Polymers. *Sens. Actuators B Chem.*, 185, 39–46, 2013.

81. Mineo, P.G., Livoti, L., Giannetto, M., Gulino, A., Schiavo, L., Cardiano, P., Very fast CO2 response and hydrophobic properties of novel poly(ionic liquid)s. *J. Mater. Chem.*, 19, 8861–70, 2009.

82. Kukla, A.L., Pavluchenko, A.S., Shirshov, Y.M., Konoshchuk, N.V., Posudievsky, O.Y., Application of sensor arrays based on thin films of conducting polymers for chemical recognition of volatile organic solvents. *Sens. Actuators B*, 135, 541–51, 2009.

83. Pegalajar, M.C., Printed Disposable Colorimetric Array for Metal Ion Discrimination. *Anal. Chem.*, 86, 8634–41, 2014.

84. Ashina, J., Kirsanov, D., Moreau, M., Koverga, V., Mikhelson, K., Ruckebusch, C. *et al.*, Raman transduction for polymeric ion-selective sensor membranes: Proof of concept study. *Sens. Actuators B Chem.*, 253, 697–702, 2017.

85. Liu, B., Dai, H., Bao, Y., Du, F., JT and, R.B., 2,6-Substituted pyridine derivative-containing conjugated polymers: Synthesis, photoluminescence and ion-sensing properties. *Polym. Chem.*, 2, 1699–705, 2011.

86. Ding, B., Si, Y., Wang, X., Yu, J., LF and, G.S., Label-free ultrasensitive colorimetric detection of copper(II) ions utilizing polyaniline/polyamide-6 nano-fiber/net sensor strips. *J. Mater. Chem.*, 21, 13345–53, 2011.

87. Fazeli, E., Pera, N., Rosenholm, J.M., Arppe, R., Soukka, T., Ratiometric Sensing and Imaging of Intracellular pH Using Polyethylenimine-Coated Photon Upconversion Nanoprobes. *Anal. Chem.*, 89, 1501–8, 2017.

88. Mariani, F., Gualandi, I., Tessarolo, M., Fraboni, B., S, E., PEDOT: Dye-Based, Flexible Organic Electrochemical Transistor for Highly Sensitive pH Monitoring. *ACS Appl. Mater. Interfaces*, 10, 22474–84, 2018.

89. Gong, J., Venkateswaran, S., Tanner, M.G., Stone, J.M., Bradley, M., Polymer Microarrays for the Discovery and Optimization of Robust Optical-Fiber-Based pH Sensors. *ACS Comb. Sci.*, 21, 417–24, 2019.

90. Miao, J., Chen, Y., Li, Y., Cheng, J., Wu, Q., Ng, K.W. *et al.*, Proton Conducting Polyoxometalate/Polypyrrole Films and Their Humidity Sensing Performance. *ACS Appl. Nano Mater.*, 1, 564–71, 2018.

91. Dai, J., Zhao, H., Lin, X., Liu, S., Liu, Y., Liu, X. *et al.*, Ultrafast Response Polyelectrolyte Humidity Sensor for Respiration Monitoring. *ACS Appl. Mater. Interfaces*, 11, 6483–90, 2019.

92. Hossein-babaei, F., Akbari, T., Harkinezhad, B., Dopant passivation by adsorbed water monomers causes high humidity sensitivity in PEDOT: PSS thin films at ppm-level humidity. *Sens. Actuators B Chem.*, 293, 329–35, 2019.

93. Zhang, J., Liu, X., Neri, G., Pinna, N., Nanostructured Materials for Room-Temperature Gas Sensors, *Advanced Mater.*, 28, 795–831, 2016.

94. Sun, Y. F., Liu, S. B., Meng, F. L., Liu, J. Y., Jin, Z., Kong, L.T., Liu, J. H., Metal Oxide Nanostructures and Their Gas Sensing Properties: A Review. *Sensors*, 12, 2610–31, 2012.

95. Das, T.K., and Prusty, S., Review on Conducting Polymers and Their Applications. *Polymer-Plastics Technology and Engineering*, 51, 1487–1500 2012.

96. Ionov, L., Polymeric Actuators, *Langmuir*, 31, 18, 5015–5024, 2015.

97. Adhikari, B. and Majumdar, S., Polymers in sensor applications. 29, 699–766, 2004.

98. Pandis, C., Peoglos, V., Kyritsis, A., Pissis, P., Gas sensing properties of conductive polymer nanocomposites, *Proc. Eng.*, 25, 243–246, 2011.

99. Shukla, S.K., Shekhar, C., Singh, N.B., Recent developments in conducting polymer based composites. *Mater. Today Proc.*, 4, 5672–81, 2017.

100. Sarkar, T., Srinives, S., Sarkar, S., Haddon, R.C., Mulchandani, A., Single-Walled Carbon Nanotube Poly(porphyrin) Hybrid for Volatile Organic Compounds Detection, *J. Phys. Chem. C*, 118, 1602–1610, 2014.

101. Li, Y., Ban, H., Yang, M., Highly sensitive NH3 gas sensors based on novel polypyrrole-coated SnO2 nanosheet nanocomposites. *Sens. Actuators B Chem.*, 224, 449–457, 2015.

102. Shukla, S.K., Kushwaha, C.S., Shukla, A., Dubey, G.C., Integrated approach for efficient humidity sensing over zinc oxide and polypyrole composite. *Mater. Sci. Eng. C*, 90, 325–332, 2018.

103. Gong, J., Li, Y., Hu, Z., Zhou, Z., Deng, Y., Ultrasensitive NH3 Gas Sensor from Polyaniline Nanograin Enchased TiO2 Fibers. *J. Phys. Chem. C*, 114, 9970–4, 2010.

104. Lu, X., Zhang, W., Wang, C., Wen, T., Wei, Y., Progress in Polymer Science One-dimensional conducting polymer nanocomposites: Synthesis, properties and applications. *Prog. Polym. Sci.*, 36, 671–712, 2011.

105. Hussain, R., Khan, M.Q., Ali, A., Electrochemical sensing of Pb2+ ion in water by ion selective membrane electrode based on Polypyrrole titanium(IV) sulphosalicylo phosphate cation exchange nanocomposite. *Groundwater Sustainable Dev.*, 8, 216–225, 2018.

106. Jarolímová, Z., Han, T., Mattinen, U., Bobacka, J., Bakker, E., Capacitive Model for Coulometric Readout of Ion-Selective Electrodes, *Anal. Chem.* 90, 8700–8707, 2018.

107. Bobacka, J., Ivaska, A., Lewenstam, A., Potentiometric Ion Sensors Based on Conducting Polymers. *Electroanalysis*, 15, 366–74, 2003.

108. Gupta, A., Singh, T.S., Yadava, R.D.S., Polymer-Coated MEMS Chemical Sensor Array for Monitoring Oxidative Stress by Breath Analysis, *International Conference on Innovations in Information, Embedded and Communication Systems* (ICIIECS), Coimbatore, pp. 1–8. 2017.

109. Liu, X., Cheng, S., Liu, H., Hu, S., Zhang, D., Ning, H. A Survey on Gas Sensing Technology. *Sensors*, 12, 9635–65, 2012.

110. Mun, S., Park, Y., Lee, Y.K., Sung, M.M., Highly Sensitive Ammonia Gas Sensor Based on Single-Crystal Poly (3-hexylthiophene) (P3HT) Organic Field Effect Transistor. *Langmuir*, 33, 13554–60, 2017.

111. Shang, Y.Y., Tin, W.T., Hong-Yu, Y., Guan-Yu, C., Che-Chi, S., Yu-Chih, L., Chang-Chiang, C., Hsiao-Wen, Z., Hsin-Fei, M., Chia-Jung, L., Chien-Lung, W., Wen-Bin, J., S, O., A Versatile Method to Enhance the Operational Current of Air-Stable Organic Gas Sensor for Monitoring of Breath Ammonia in Hemodialysis Patients. *ACS Sens.*, 4, 1023–31, 2019.

112. Zhang, D., Fan, X., Hao, X., Dong, G., Facile Fabrication of Polyaniline Nanocapsule Modified Zinc Oxide Hexagonal Microdiscs for H2S Gas Sensing Applications. *Ind. Eng. Chem. Res.*, 58, 1906–13, 2019.

113. Han, S., Yang, Z., Li, Z., Zhuang., X., Akinwande, D., Yu, J., Improved Room Temperature NO2 Sensing Performance of Organic Field-Effect Transistor by Directly Blending a Hole-Transporting/Electron-Blocking Polymer into the Active Layer. *ACS Appl. Mater. Interfaces*, 10, 38280–6, 2018.

114. Al-mashat, L., Debiemme-chouvy, C., Borensztajn, S., Wlodarski, W., Electropolymerized Polypyrrole Nanowires for Hydrogen Gas Sensing. *J. Phys. Chem. C*, 116, 13388–13394, 2012.

115. Hong, J., Lee, S., Seo, J., Pyo, S., Kim, J., Lee, T.A., Highly Sensitive Hydrogen Sensor with Gas Selectivity Using a PMMA Membrane-Coated Pd Nano-particle/Single-Layer Graphene Hybrid. *ACS Appl. Mater. Interfaces*, 7, 3554–3561, 2015.

116. Ma, Z., Chen, P., Cheng, W., Yan, K., Pan, L., Shi, Y. *et al.*, Highly Sensitive, Printable Nanostructured Conductive Polymer Wireless Sensor for Food Spoilage Detection. *Nano Lett.*, 18, 4570–5, 2018.

117. Kumar, V., Patil, V., Apte, A., Harale, N., Patil, P., Kulkarni, S., Ultrasensitive Gold Nanostar – Polyaniline Composite for Ammonia Gas Sensing. *Langmuir*, 31, 13247–13256, 2015.

118. Wang, F., Wang, Y., Yu, H., Chen, J., Gao, B., Lang, J., One Unique 1D Silver(I)-Bromide-Thiol Coordination Polymer Used for Highly Efficient Chemiresistive Sensing of Ammonia and Amines in Water. *Inorg. Chem.*, 55, 9417–9423, 2016.

119. Wu, H., Chen, Z., Zhang, J., Wu, F., He, C., Ren, Z. Wu, Y., Manipulating Polyaniline Fibrous Networks by Doping Tetra- ß – carboxyphthalocyanine Cobalt(II) for Remarkably Enhanced Ammonia Sensing. *Chem. Mater.*, 29, 9509–17, 2017.

120. Yu, S.H., Cho, J., Sim, K.M., Ha, J.U., Chung, D.S., Morphology-Driven High-Performance Polymer Transistor-based Ammonia Gas Sensor. *ACS Appl. Mater. Interfaces*, 8, 6570–6576, 2016.

122. Yang, Y., Zhang, G., Luo, H., Yao, J., Liu, Z., Zhang, D., Highly Sensitive Thin-Film Field-Effect Transistor Sensor for Ammonia with the DPP-Bithiophene Conjugated Polymer Entailing Thermally Cleavable tert -Butoxy Groups in the Side Chains. *ACS Appl. Mater. Interfaces*, 8, 3635–3643, 2016.

123. Menegazzo, N., Boyne, D., Bui, H., Beebe, T.P., Booksh, K.S., DC Magnetron Sputtered Polyaniline-HCl Thin Films for Chemical Sensing Applications. *Anal. Chem.*, 84, 5770–5777, 2012.

124. Kanawade, R., Kumar, A., Pawar, D., Late, D., Mondal, S., Sinha, R.K., Fiber optic Fabry–Perot interferometer sensor: An efficient and fast approach for ammonia gas sensing. *J. Opt. Soc. Am. B*, 36, 684–9, 2019.

125. Shu, J., Qiu, Z., Lv, S., Zhang, K., Tang, D., Cu2+ -Doped SnO2 Nanograin/ Polypyrrole Nanospheres with Synergic Enhanced Properties for Ultra-sensitive Room-Temperature H2S Gas Sensing. *Anal. Chem.*, 89, 11135–42, 2017.

126. Kitture, R., Pawar, D., Rao, C., Choubey, R., Kale, S., Nanocomposite modified optical fiber: A room temperature, selective H2S gas sensor: Studies using ZnO-PMMA. *J. Alloys Compd.*, 695, 2091–6, 2017.

127. Bai, S., Guo, J., Sun, J., Tang, P., Chen, A., Luo, R., Li, D., of NO2 Sensing Performance at Room Temperature by Graphene-Modified Polythiophene. *Ind. Eng. Chem. Res.*, 55, 5788–5794, 2016.

128. Li, H., Dailey, J., Kale, T., Besar, K., Koehler, K., Katz, H.E., Sensitive and Selective NO2 Sensing Based on Alkyl- and Alkylthio- Thiophene Polymer Conductance and Conductance Ratio Changes from Differential Chemical Doping. *ACS Appl. Mater. Interfaces*, 9, 20501–20507, 2017.

129. Yuan, W., Huang, L., Zhou, Q., Shi, G., Ultrasensitive and Selective Nitrogen Dioxide Sensor Based on Self-Assembled Graphene/Polymer Composite Nano fi bers. *ACS Appl. Mater. Interfaces*, 6, 17003–17008, 2014.

130. Al-mashat, L., Shin, K., Kalantar-zadeh, K., Plessis, J.D., Han, S.H., Kojima, R.W. Kaner , R. B., Li, D., Gou, X., Ippolito, S., Wlodarski, W., Graphene/ Polyaniline Nanocomposite for Hydrogen Sensing., *J. Phys. Chem. C*, 114, 16168–73, 2010.

131. Won, S.C., Chang, S.L., Long, H., Myoung, H.O., Alex, Z., Carlo, C., Jong, H.K., M, R., Direct Organization of Morphology-Controllable Mesoporous SnO2 Using Amphiphilic Graft Copolymer for Gas-Sensing Applications. *ACS Appl. Mater. Interfaces*, 9, 37246–53, 2017.

132. Barauskas, D., Park, S.J., Pelenis, D., Vanagas, G., Lee, J.J., Virzonis, D., Jones, C.W.., Baltrusaitis, J., CO2 and SO2 Interactions with Methylated Poly(ethylenimine)-Functionalized Capacitive Micromachined Ultrasonic Transducers (CMUTs): Gas Sensing and Degradation Mechanism. *ACS Appl. Electron. Mater.*, 1, 1150–1161, 2019.

133. Chang, Y., Hasan, D., Dong, B., Wei, J., Ma, Y., Zhou, G. *et al.*, All-Dielectric Surface-Enhanced Infrared Absorption-Based Gas Sensor Using Guided Resonance. *ACS Appl. Mater. Interfaces*, 10, 38272–9, 2018.

134. Promthaveepong, K. and Li, N., CO2 Responsive Polymer-Functionalized Au Nanoparticles for CO2 Sensor. *Anal. Chem.*, 88, 8289–8293, 2016.

135. Yoon, B., Choi, S., Swager, T.M., Walsh, G.F., Switchable Single-Walled Carbon Nanotube–Polymer Composites for CO2 Sensing. *ACS Appl. Mater. Interfaces*, 10, 33373–9, 2018.

136. Kochmann, S., Baleiza, C., Berberan-santos, N., Wolfbeis, O.S., Sensing and Imaging of Oxygen with Parts per Billion Limits of Detection and Based on the Quenching of the Delayed Fluorescence of 13C70 Fullerene in Polymer Hosts. *Anal. Chem.*, 85, 1300–4, 2013.

137. Muckley, E.S., Collins, L., Ievlev, A.V., Ye, X., Kisslinger, K., Sumpter, B.G. *et al.*, Light-Activated Hybrid Nanocomposite Film for Water and Oxygen Sensing. *ACS Appl. Mater. Interfaces*, 10, 31745–54, 2018.

138. Tolentino M., Albano, D., Sevilla III, F., Piezoelectric Sensor for Ethylene based on Silver(I)/Polymer Composite. *Sens. Actuators B Chem.*, 254, 299–306, 2017.

139. Barkade, S.S., Pinjari, D.V., Singh, A.K., Gogate, P.R., Naik, J.B., Sonawane, S.H. *et al.*, Ultrasound Assisted Miniemulsion Polymerization for Preparation of Polypyrrole – Zinc Oxide (PPy/ZnO) Functional Latex for Liquefied Petroleum Gas Sensing. *Ind. Eng. Chem. Res.*, 52, 7704–7712, 2013.

140. Klini, A., Pissadakis, S., Das, R.N., Giannelis, E.P., Anastasiadis, S.H., Anglos, D., ZnO – PDMS Nanohybrids: A Novel Optical Sensing Platform for Ethanol Vapor Detection at Room Temperature. *J. Phys. Chem. C*, 119, 623–31, 2015.

141. Lova, P., Manfredi, G., Boarino, L., Comite, A., Laus, M., Patrini, M, Marabelli, F., Soci, C., Comoretto, D., Polymer Distributed Bragg Reflectors for Vapor Sensing. *ACS Photonics*, 2, 537–543, 2015.

142. Wang, X., Hou, S., Goktas, H., Kovacik, P., Yaul, F., Paidimarri, A. *et al.*, Small-Area, Resistive Volatile Organic Compound (VOC) Sensors Using Metal – Polymer Hybrid Film Based on Oxidative Chemical Vapor Deposition (oCVD). *ACS Appl. Mater. Interfaces*, 7, 16213–16222, 2015.

143. Campos, R., Reuther, J.F., Mammoottil, N.R., Novak, B.M., Solid State Sensing of Nonpolar VOCs Using the Bistable Expansion and Contraction of Helical Polycarbodiimides. *Macromolecules*, 50, 4927–4934, 2017.

144. Sudeshna, B., Rashmi, A., Mondal, S.K., Electrospun polypyrrole-polyethylene Oxide coated Optical fiber Sensor probe for detection Of volatile compounds. *Sens. Actuators B Chem.*, 250, 52–60, 2017.

145. Li, F., Li, H., Jiang, H., Zhang, K., Chang, K., Jia, S., Jiang, W., Shang, Y., Lu, W., Deng, S., Chen, M. Polypyrrole nanoparticles fabricated via Triton X-100 micelles template approach and their acetone gas sensing property. *Appl. Surf. Sci.*, 280, 212–218, 2013.

146. Mondal, R.K., Dubey, K.A., Kumar, J., Bhardwaj, Y.K., Melo, J.S., Varshney, L., Carbon Nanotube Functionalization and Radiation Induced Enhancements in the Sensitivity of Standalone Chemiresistors for Sensing Volatile Organic Compounds. *ACS Appl. Nano Mater.*, 1, 5470–82, 2018.

147. Sachdeva, S., Koper, S.J.H., Sabetghadam, A., Soccol, D., Gravesteijn, D.J., Kapteijn, F. *et al.*, Gas Phase Sensing of Alcohols by Metal Organic Framework – Polymer Composite Materials. *ACS Appl. Mater. Interfaces*, 9, 24926–24935, 2017.

148. Iftekhar, U.A.S.M., Usman, Y., C., G.-S., Improving the Working E ffi ciency of a Triboelectric Nanogenerator by the Semimetallic PEDOT: PSS Hole Transport Layer and Its Application in Self-Powered Active Acetylene Gas Sensing. *ACS Appl. Mater. Interfaces*, 8, 30079–30089, 2016.

149. Bastakoti, B.P., Torad, N.L., Yamauchi, Y., Polymeric Micelle Assembly for the Direct Synthesis of Platinum- Decorated Mesoporous TiO2 toward Highly Selective Sensing of Acetaldehyde. *ACS Appl. Mater. Interfaces*, 6, 854–860, 2014.

150. Hou, K., Rehman, A., Zeng, X., Study of Ionic Liquid Immobilization on Polyvinyl Ferrocene Substrates for Gas Sensor Arrays. *Langmuir*, 27, 5136–46, 2011.

151. Truax, S.B., Demirci, K.S., Beardslee, L.A., Luzinova, Y., Hierlemann, A., Mizaiko, B. *et al.*, Mass-Sensitive Detection of Gas-Phase Volatile Organics Using Disk Microresonators. *Anal. Chem.*, 83, 3305–11, 2011.

152. Susmita, D., Susanta, K.B., Stella, S.B., Sofiya, K., Leila, Z., J, R., Colorimetric Polydiacetylene – Aerogel Detector for Volatile Organic Compounds (VOCs). *ACS Appl. Mater. Interfaces*, 9, 2891–2898, 2017.

153. Ngo, Y.H., Brothers, M., Martin, J.A., Grigsby, C.C., Fullerton, K., Naik, R.R. *et al.*, Chemically Enhanced Polymer-Coated Carbon Nanotube Electronic Gas Sensor for Isopropyl Alcohol Detection. *ACS Omega*, 3, 6230–6, 2018.

154. Kanawade, R., Kumar, A., Pawar, D., Vairagi, Late, Mondal, Sinha, Negative axicon tip-based fiber optic interferometer cavity sensor for volatile gas sensing. *Opt. Express*, 27, 7277–90, 2019.

155. Wei, L., Jiawei, Z., Youju, H., Patrick, T., Qing, H., C., T., Self-Diffusion Driven Ultrafast Detection of ppm-Level Nitroaromatic Pollutants in

Aqueous Media Using a Hydrophilic Fluorescent Paper Sensor. *ACS Appl. Mater. Interfaces*, 9, 23884–23893, 2017.

156. Yuqi, Z., Yimin, S., Jiaqi, L., Pu, G., Zhongyu, C.J.J.W., Polymer-infiltrated SiO2 inverse opal photonic crystals for colorimetrically selective detection of xylene vapors. *Sens. Actuators B Chem.*, 291, 67–73, 2019.

157. Makkad, S.K. and Sk, A., Surface Functionalized Fluorescent PS Nanobead Based Dual-Distinct Solid State Sensor for Detection of Volatile Organic Compounds. *Anal. Chem.*, 90, 7434–41, 2018.

158. Gopalakrishnan, D. and Dichtel, W.R., Direct Detection of RDX Vapor Using a Conjugated Polymer Network. *J. Am. Chem. Soc.*, 135, 8357–8362, 2013.

159. Wang, X., Ugur, A., Goktas, H., Chen, N., Wang, M., Lachman, N. *et al.*, Room Temperature Resistive Volatile Organic Compound Sensing Materials Based on a Hybrid Structure of Vertically Aligned Carbon Nanotubes and Conformal oCVD/iCVD Polymer Coatings. *ACS Sens.*, 1, 374–383, 2016.

160. Kim, J., Smartphone-Based VOC Sensor Using Colorimetric Polydiace-tylenes. *Appl. Mater. Interfaces*, 10, 5014–21, 2018.

161. Ding, W., Xu, J., Wen, Y., Zhang, J., Liu, H., Zhang, Z., Highly selective "turn-on" fluorescent sensing of fluoride ion based on a conjugated polymer thin film-Fe3+ complex. *Anal. Chim. Acta*, 967, 78–84. 2017.

162. Kim, S.K., Gupta, M., Lee, H., A recyclable polymeric film for the consecutive colorimetric detection of cysteine and mercury ions in the aqueous solution. *Sens. Actuators B Chem.*, 257, 728–33, 2018.

163. Gupta, M. and Lee, H., Dyes and Pigments Water-soluble polymeric probe with dual recognition sites for the sequential colorimetric detection of cyanide and Fe (III) ions. *Dyes Pigm.*, 167, 174–80, 2019.

164. Cheng, Q., Lan, J., Chen, Y., Lin, J., Chenna, R., Reddy, K. *et al.*, 1D helical silver(I)-based coordination polymer containing pyridydiimide ligand for Fe(III) ions detection. *Inorg. Chem. Commun.*, 96, 30–33, 2018.

165. Chen, Z., Mi, X., Wang, S., Lu, J., Li, Y., Li, D., Dou, J., Two Novel Penetrating Coordination Polymers Based on Flexible S-containing dicarboxylate acid with Sensing Properties Towards Fe3+ and Cr2O7 2- ions. *J. Solid State Chem.*, 261, 75–85, 2018.

166. Hu, Y., Gao, Z., Yang, J., Chen, H., Han, L., Environmentally benign conversion of waste polyethylene terephthalate to fluorescent carbon dots for "on-off-on" sensing of ferric and pyrophosphate ions. *J. Colloid Interface Sci.*, 538, 481–488, 2018.

167. Huang, Y., Gao, L., Zhang, J., Wang, X., Fan, M., Hu, T., Two novel luminescent Cd(II)/Zn(II) coordination polymers Based on 4,4'-(1H-1,2,4-triazol1yl)methylene bis(benzonic acid) For sensing organic Molecules and Fe3+ ion. *Inorg. Chem. Commun.*, 91, 35–38, 2018.

168. Kamac, M. and Kaya, İ., Polymeric fluorescent film sensor based on poly (azomethine-urethane): Ion sensing and surface properties. *React. Funct. Polym.*, 136, 1–8, 2019.

169. Hande, P.E., Samui, A.B., Kulkarni, P.S., Selective Nanomolar Detection of Mercury using Coumarin based Fluorescent Hg (II) -Ion Imprinted Polymer. *Sens. Actuators B Chem.*, 246, 597–605, 2017.

170. Jia, Y., Pan, Y., Wang, H., Chen, R., Wang, H., Cheng, X., Highly selective and sensitive polymers with fluorescent side groups for the detection of Hg 2 þ ion. *Mater. Chem. Phys.*, 196, 262–9, 2017.

171. Avudaiappan, G., AJ, K., Theresa, L.V., Shebitha, A.M., Hiba, K., Shenoi, P.K., A novel dendritic polymer based turn- off fluorescence sensor for the selective detection of cyanide ion in aqueous medium. *React. Funct. Polym.*, 137, 71–8, 2019.

172. Chen, L., Weng, N., Wu, W., Chen, W., Electrospun polymer nanofibers of P(NIPAAm-co-SA-co-FBPY): Preparation, structural control, metal ion sensing and thermoresponsive characteristics. *Mater. Chem. Phys.*, 163, 63–72, 2015.

173. Aki, S., Maeno, K., Sueyoshi, K., Hisamoto, H., Endo, T. Development Of A Polymer/Tio2 Hybrid Two-Dimensional Photonic Crystal For Highly Sensitive Fluorescence-Based Ion Sensing Applications. *Sens. Actuat. B Chem*, 269, 257–263, 2018.

174. Alahi, E.E., Mukhopadhyay, S.C., Burkitt, L. Imprinted polymer coated impedimetric nitrate sensor for real- time water quality monitoring. *Sens. Actuat. B Chem*, 259, 753–761, 2017.

175. Boali, A.A., Mansha, M., Waheed, A., Ullah, N. Synthesis and selective colorimetric detection of iodide ion by novel 1,5-naphthyridine-base d conjugate d polymers. *J Taiwan Inst Chem Eng.* 91, 420–426, 2018.

176. Dahaghin, Z., Kilmartin, P.A., Mousavi, H.Z., Determination of cadmium(II) using a glassy carbon electrode modified with a Cd-ion imprinted polymer. *J. Electroanal. Chem.*, 810, 185–190, 2018.

177. Kan, W. and Wen, S., A fluorescent coordination polymer for selective sensing of hazardous nitrobenzene and dichromate anion. *Dyes Pigm.*, 139, 372–80, 2017.

178. Kim, Y., Jang, G., Lee, T. S., Carbon nanodots functionalized with rhodamine and poly(ethylene glycol) for ratiometric sensing of Al ions in aqueous solution, *Sens. Actuat. B Chem.*, 249, 59–65, 2017.

179. Popat, A., Kumeria, T., Santos, A., Environmental Copper Sensor Based on Polyethylenimine-Functionalized Nanoporous Anodic Alumina Interferometers. *Anal. Chem.*, 91, 5011–20, 2019.

180. Frankær, C.G., Hussain, K.J., Doürge, T.C., S, T.J., Optical Chemical Sensor Using Intensity Ratiometric Fluorescence Signals for Fast and Reliable pH Determination. *ACS Sens.*, 4, 4–9, 2019.

181. Duong, H.D., Shin, Y., Rhee, J.I., Development of fluorescent pH sensors based on a sol-gel matrix for acidic and neutral pH ranges in a microtiter plate. *Microchem. J.*, 147, 286–95, 2019.

182. Patil, S., Ghadi, H., Ramgir, N., Adhikari, A., Rao, V.R., Monitoring soil pH variation using Polyaniline/SU-8 composite film based conductometric microsensor. *Sens. Actuators B Chem.*, 286, 583–90, 2019.

183. Dang, W., Manjakkal, L., Navaraj, W.T., Lorenzelli, L., Vinciguerra, V., Dahiya, R., Stretchable wireless system for sweat pH monitoring. *Biosens. Bioelectron.*, 107, 192–202, 2018.

184. Zhao, Y., Lei, M., Liu, S.X., Zhao, Q., Smart hydrogel-based optical fiber SPR sensor for pH measurements. *Sens. Actuators B Chem.*, 261, 226–32, 2018.

185. Shi, X., Mei, L., Zhang, N., Zhao, W., Xu, J., Chen, H.A., Polymer Dots-Based Photoelectrochemical pH Sensor: Simplicity, High Sensitivity, and Broad-Range pH Measurement. *Anal. Chem.*, 90, 8300–3, 2018.

186. Al-Qaysi, W.W. and Duerkop, A., Sensor and sensor microtiterplate with expanded pH detection range and their use in real samples. *Sens. Actuators B Chem.*, 298, 126848, 2019.

187. Hegarty, C., McConville, A., McGlynn, R.J., Mariotti, D., Davis, J., Design of composite microneedle sensor systems for the measurement of transdermal pH. *Mater. Chem. Phys.*, 227, 340–6, 2019.

188. Abdollahi, A., Mouraki, A., Sharifian, M.H., Mahdavian, A.R., Photochromic properties of stimuli-responsive cellulosic papers modified by spiropyran-acrylic copolymer in reusable pH-sensors. *Carbohydr. Polym.*, 200, 583–94, 2018.

189. Zuaznabar, G.J.C. and Fragoso, A.A., wide-range solid state potentiometric pH sensor based on poly-dopamine coated carbon nano-onion electrodes. *Sens. Actuators B Chem.*, 273, 664–71, 2018.

190. Deng, K., Bellmann, C., Fu, Y., Rohn, M., Guenther, M., Gerlach, G., Miniaturized force-compensated hydrogel-based pH sensors. *Sens. Actuators B Chem.*, 255, 3495–504, 2018.

191. Su, F., Agarwal, S., Pan, T., Qiao, Y., Zhang, L., Shi, Z. *et al.*, Multifunctional PHPMA-Derived Polymer for Ratiometric pH Sensing, Fluorescence Imaging, and Magnetic Resonance Imaging. *Appl. Mater. Interfaces*, 10, 1556–65, 2018.

192. Mallick, S., Ahmad, Z., Touati, F., Shakoor, R.A., Improvement of humidity sensing properties of PVDF-TiO2 nanocomposite films using acetone etching. *Sens. Actuat. B Chem.*, 288, 408–413, 2019.

193. Zhang, Y. and Cui, Y., Flexible Calligraphy-Integrated In-situ Humidity Sensor. *Measurement*, 147, 106853, 2019.

194. Duy, L.V., Yi-Ying, L., Ting-Han, L., Ming-Chung, W., Fabrication and humidity sensing property of UV/ozone treated PANI/PMMA electrospun fibers. *J. Taiwan Inst. Chem. Eng.*, 99, 250–257, 2019.

195. Chethan, B., Prakash, H.G.R., Ravikiran, Y.T., Vijayakumari, S.C., Thomas, S., Polypyyrole based core-shell structured composite based humidity Sensor operable at room temperature. *Sens. Actuators B Chem.*, 296, 126639, 2019.

196. Dehghani, M., Esmailzadeh, F., Bahrampour, A., A proposal for distributed humidity sensor based on the induced LPFG in a periodic polymer coated fiber structure. *Opt. Laser Technol.*, 117, 126–33, 2019.

197. Ru, C., Gu, Y., Li, Z., Duan, Y., Zhuang, Z., Na, H. *et al.*, Effective enhancement on humidity sensing characteristics of sulfonated poly (ether ether ketone) via incorporating a novel bifunctional metal – organic – framework. *J. Electroanal. Chem.*, 833, 418–26, 2019.

198. Kang, T., Park, J., Kim, B., Lee, J.J., Choi, H., Lee, H., Yook, J., Microwave characterization of conducting polymer PEDOT: PSS film using a microstrip line for humidity sensor application. *Measurement*, 137, 272–277, 2019.

199. Trchov, M. and Stejskal, J., Microcomposites of zirconium phosphonates with a conducting polymer, polyaniline: Preparation, spectroscopic study and humidity sensing. *J. Solid State Chem.*, 276, 285–93, 2019.

200. Zhang, D., Chen, H., Zhou, X., Wang, D., Jin, Y., *In-situ* polymerization of metal organic frameworks-derived ZnCo 2 O 4/polypyrrole nanofilm on QCM electrodes for ultra-highly sensitive humidity sensing application. *Sens. Actuators A Phys.*, 295, 687–95, 2019.

201. Wei, X., X, C., Feifei, Q., S, Y., Z, Z., Ye, Z., Whispering-gallery mode lasing from polymer microsphere for humidity sensing. *Chin. Opt. Lett.*, 16, 6–10, 2018.

202. Ko, J., Yoon, Y., Lee, J. Quartz Tuning Forks with Hydrogel Patterned by Dynamic Mask Lithography for Humidity Sensing. *Sens. Actuat. B Chem*, 273, 821–825, 2018.

203. Anju, V.P., Jithesh, P.R., Narayanankutty, S.K. A Novel Humidity and Ammonia Sensor Based on Nanofibers/Polyaniline/Polyvinyl Alcohol. *Sens. Actuat. A Phys.*, 285, 35–44, 2018.

204. Jiang, K., Fei, T., Zhang, T. Humidity sensing properties of LiCl-loaded porous polymers with good stability and rapid response and recovery. *Sens. Actuat. B Chem.*, 199, 1–6, 2014.

205. Ranjani, L.S., Ramya, K., Dhathathreyan, K.S., Compact and flexible hydrocarbon polymer sensor for sensing humidity in confined spaces. *Int. J. Hydrogen Energy*, 39, 21343–50, 2014.

206. Wang, P., Gu, F., Zhang, L., Tong, L., Polymer microfiber rings for high-sensitivity optical humidity sensing. 50, 7–10, 2011.

207. Huang, X., Li, B., Wang, L., Lai, X., Xue, H., Gao, J., Superhydrophilic, Underwater Superoleophobic, and Highly Stretchable Humidity and Chemical Vapor Sensors for Human Breath Detection. *ACS Appl. Mater. Interfaces*, 11, 24533–43, 2019.

208. Wu, J., Sun, Y., Wu, Z., Li, X., Wang, N., Tao, K. *et al.*, Carbon Nanocoil-Based Fast-Response and Flexible Humidity Sensor for Multifunctional Applications. *ACS Appl. Mater. Interfaces*, 11, 4242–51, 2019.

209. Su, C., Chiu, H., Chen, Y., Yesilmen, M., Schulz, F., Ketelsen, B. *et al.*, Highly Responsive PEG/Gold Nanoparticle Thin-Film Humidity Sensor via Inkjet Printing Technology. *Langmuir*, 35, 3256–64, 2019.

210. Kou, D., Ma, W., Zhang, S., Lutkenhaus, J.L., Tang, B., High-Performance and Multifunctional Colorimetric Humidity Sensors Based on Mesoporous

Photonic Crystals and Nanogels. *ACS Appl. Mater. Interfaces*, 10, 41645–54, 2018.

211. Lee, C., You, Y., Dai, J., Hsu, J., Horng, J., Hygroscopic polymer microcavity fiber Fizeau interferometer incorporating a fiber Bragg grating for simultaneously sensing humidity and temperature. *Sens. Actuators B Chem.*, 222, 339, 2016.

6

Shape Memory Actuators

Sithara Gopinath[1], Dr. Suresh Mathew[2*] and Dr. P. Radhakrishnan Nair[1]

[1]*Advanced Molecular Materials Research Centre (AMMRC),*
Mahatma Gandhi University, Kottayam, India
[2]*School of Chemical Sciences, Mahatma Gandhi University, Kottayam, India*

Abstract

Actuators are nano-, micro-, and macroscale working mechanical devices that can change their shape with respect to the environmental conditions. Shape memory actuators are actuators that are made from materials that exhibit shape memory property or shape memory effect. It is a distinctive effect, which can return to an original geometry or permanent shape, after a large inelastic deformation (near 10%). Shape memory actuators mainly adopted in applications where huge power and stroke are needed. After Ni-Ti alloys shape memory innovation, there occurred fast progress in the various areas of applications, such as medical field, electronics, avionics, mechanical automobiles, and consumer, as well as biomedical products. These materials or devices have dimensions varying from micrometers range to meters. The present chapter deals with recent emerging trends in the area of these shape memory actuators, its classifications, characteristics, applications, etc.

Keywords: Shape change effect (SCE), glass transition temperatures, polymer actuators, shape memory alloys (SMAs), shape memory technology (SMT), fatigue

6.1 Introduction

Basically, actuators are devices that work on the basis of motion and regulating a process or any systems. It can be also defined as something that converts energy into motion. SME can be elucidated as a novel and amazing process of retaking or regaining the original shape from its already

Corresponding author: sureshmathewmgu@gmail.com

Inamuddin, Rajender Boddula and Abdullah M. Asiri (eds.) Actuators: Fundamentals, Principles, Materials and Applications, (139–158) © 2020 Scrivener Publishing LLC

programmed or predesigned transformed shape, with the application of suitable stimulus [1, 2] given externally as either heat, light, electric, or magnetic fields, etc. Based on the conventional understanding from the works of literature, the word "SME" strikes the attention in 1932, from the researches on amazing shape recovery phenomenon in Au-Cd alloy [3, 4]. This SME effect was utilized in real-time engineering applications already before 1906, considerably prior to temperature-recoil polymer made materials began to utilize for insulating cable bundles [5]. Shape memory materials (SMMs) are materials that exhibit remarkable SMEs [2]. As researches going on, there is an amazing expansion in the number of actuators from metallic alloys such as Ni-Ti (Nickel-Titanium) and Cu-Al-Ni (Copper-Aluminium-Nickel) [6] and polymers (which includes gels and liquid crystals) to hybrids and ceramics, etc. [5, 7–9]. The classification of these materials on SME is strongly based on the applied external stimuli used for triggering purpose: i.e., thermo-responsive (*via* thermal stimulus), photo-responsive (*via* light stimulus), chemo-responsive (*via* solvents or chemicals), and magneto-responsive (*via* magnetic field) [2, 5, 8, 10]. SME is sometimes confused with SCE, i.e., the shape change effect or also known as shape-changing effect which is also a shape switching process [5, 8, 10]. While the SME requires the use of the correct trigger for shape retrieval, SCE gradually regains its original effect as dismissal of correct external stimulus progresses. SME and its similar term SCE mainly differ in the energy levels of permanent or original shape and programmed or temporary shapes. Applications of SME are promising because sometimes it is very difficult to do the same by conventional materials/technologies. In 1990, the SMM community introduced shape memory technology (SMT) [11, 12]. This led to the wide use of reshaped product design in many ways [7–11, 13, 14]. We can call SMM as "the material is the machine" [12, 15] because it can integrate sensing and actuation functions together.

Figure 6.1 shows schematic diagram shape memory process. This cycle consists of two stages. The first one is a hot stage which is of off-red background and the second is a cold stage which is of a blue background. In the hot phase, shape change occurs. At first, a fixed shaped film is subjected to heat and reshaped to a corrugated ring shape with the application of temperature (external force). This is the temporary shape that is fixed after cooling. Also, we can call this stage as a programming stage, resulted in the formation of temporary or programmed shape. The final programmed material can retain its original/

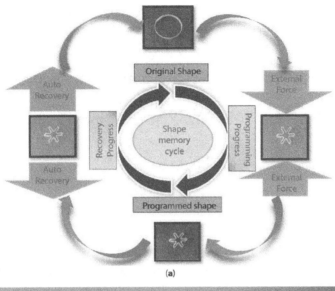

(a)

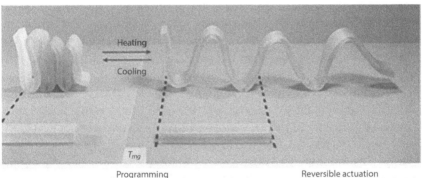

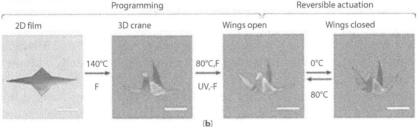

(b)

Figure 6.1 (a) Shape memory effect cycle. (b) Morphologies of shape memory polymers. Specific morphologies can be designed to enhance the function of shape memory polymers. (a) | The formation of a zig-zag pattern enables the amplification of the movement of the polymers, relative to a rectangular film. In this shape, a relatively small change in angle (below 10°) leads to a large increase in length. (b) | Origami folding enables 2D-to-3D transformation, for example, the fabrication and reversible actuation of a crane-shaped device. Programming of an active region around a fold enables the wings to open and close in response to changes in temperature (T).

permanent shape with the application of temperature. This stage is the recovered stage the process behind this is called the recovery process.

6.2   Classification of Shape Memory Actuators

Depending on the materials (SMMs) used for actuator application, we can classify them into two:

- Shape memory alloy (SMA) actuators.
- Shape memory polymer (SMP) actuators.

6.2.1   Shape Memory Alloy Actuators

As a result of certain treatment in shape memory alloys, it exhibits some temperature depending on shape change, which depends on a thermo-elastic martensitic transformation. The thermo-elastic martensitic transformation is the transformation occurred in such a way that when the applied temperature is lowered, the formed martensite stage began to grow vigorously and vice versa when the temperature is raised. This reversibility of martensite plates is mainly due to the fact that the transformation of the crystal structure is associated with very small elastic strains only [16]. The shape memory research on this is still progressing and mainly three combinations of metallic alloys which are technically applicable till now: NiTi [17, 18], Cu-Zn-Al (Copper-Zinc-Aluminium), and Cu-Al-Ni (Copper-Aluminium-Nickel).

These shape memory metallic alloys are widely used in many industrial applications due to their robust, efficient, controllable, simple, reusable, and low-density elements. The metallic compound from Nickel and Titanium (which resulted in the formation of NiTi alloy or Nitinol) along with their combinational derivatives are widely accepted as a standard alloy for 99% of shape memory actuator and also in multiple applications at the initial stage of research. Even though Nitinol possesses amazing applications, they urgently needed some major efficient modifications such as:

(i)   It cannot make use of them in automobile applications, because they lack higher transformation temperatures (M_f temperature > 80°C).

(ii)   For improving the actuator controllability, the non-linearity problem effect occurring in their strategic study of electrical resistivity must be considered.

(iii) The temperature v/s strain behavior of the obtained actuator sometimes shows wide hysteresis, which always makes a significant effect on the properties of dynamics and controllability.

Since the modifications which are listed above mainly not focuse on alloy composition, we can clearly visible a persistent attempt to improve these related properties. These researches result in tremendous potential in the field of thermomechanical improvements on the actuator material to design. In order to design a better system of actuator in a promising method, we have to make use of existing effective materials after optimization in order to make use it in required applications.

6.2.1.1   Types of Shape Memory Effect

6.2.1.1.1   One-Way SME

Consider a shape memory alloy in a martensitic state and it is then allowed to deform below a particular value; thus, there occurs deformation which can be reverted due to the motion of highly motile borders (dual borders or junction). As heat applied to the material increased, they form austenite stage which is in crystalline nature which holds a basic orientation and hence material reverted to previous shape. On subsequent cooling, there does not occur any other change in its shape and this is conventionally

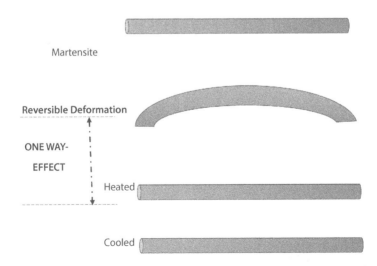

Figure 6.2 One-way shape memory effect.

called one-way SME. A schematic diagram of this effect is explained using Figure 6.2 [16, 18].

If we deform again the material in the martensitic state in the same way, it can again show the one-way SME because it is reversible.

6.2.1.1.2   Two-Way SME

If a material possesses a two-way SME, they have the ability to memorize its shape in high and low temperatures variance. For converting an alloy material to shape memory class and two-way SME, it requires a thermal and mechanical treatment. A heavy deformation in the martensite state makes the ideal way to produce two-way SME, explained with Figure 6.3 [19–21].

In this, the shape change completely depends on the amount of heat applied. When the shape change of the material in the first stage surpasses a particular temperature range, there occurs reversibly as well as not revertible shape change due to the measurements of disruptions occurred inside it. Further application of temperature on the material again is unable to recover its primary shape. But this irreversible shape change will make disturbance in the material structure. The materials reach its original shape (martensite), and thus, the third heated structure forms. Upon cooling the

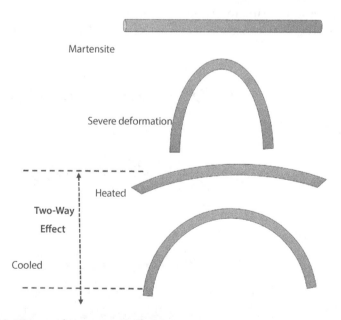

Figure 6.3 Two-way shape memory effect.

material again, martensite stage material gave the final shape because martensite plates will allocate the induced dislocation structure's stress field. This type of two-way effect is widely used in NiTi alloys because of its high ductile nature.

6.2.1.2 Basics About SMA Actuators

The SMA actuator business mainly depends on one or more elemental metals or its compounds of Nickel-Titanium along with their different phase derivatives. Due to the reduced shape memory nature and complications in the production of wires during hot and cold working, copper-based alloys such as Cu-Zn-Al, Cu-Al-Ni and Fe-based shape memory combinational alloys can be widely used.

The three main components of the SMA actuators are:

- Nitinol part for actuation (linear or coiled shape)
- Bias force
- A regulating segment used to connect electromechanical parts.

Nitinol actuators require some training procedure so that the wire would be capable to memorize the "trained" shape even after cooling by itself (Intrinsic Two-Way Effect). Along with this training procedure, a bias force also recommended, which leads to a stressful situation on the alloy material.

Nitinol actuator designs can be differentiated into three different groups. Straight Nitinol wire actuators were used for making the majority (over 90%) of electrically controlled actuators, helix-shaped compact springs were widely used for the production of different actuators heated *via* atmospheric temperature variations. Rest numbers of actuators resemble actuator calottes. Figures 6.4a–c show the three actuator types. Based on the design and electrical/mechanical requirements, the one or the other design is commonly preferable and also the selection of actuator type is taken mainly the behavior and nature of the projected application.

Important characteristics are listed below:

- Selected or relatively narrow temperature range (hysteresis) is used for the performance of the mechanical systems.
- Shape memory actuators possess a wide range of shape changes such as contraction, torsion, elongation, bending, etc.

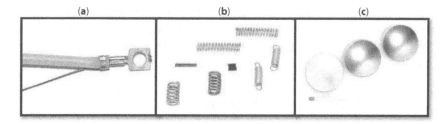

Figure 6.4 Comparison between various actuator designs. (a) shows a crimped Nitinol wire (SmartFlex 030), attached to a ring terminator using a leadcontact wire. (b) shows different spring shaped actuators for indirect heating. (c) shows different kinds of high-force-actuators which generate high force during indirect heating.

- The work density for mechano-electrical is always found to be higher.
- The shape memorizing power is bounded only to some portions of material.

6.2.1.3   Technically Suitable Shape Memory Alloys and Their Properties

Recent studies show that Ni-Ti, Cu-Zn-Al, and Cu-Al-Ni are widely used technically suitable shape memory alloys [21–23] as shown in Table 6.1. The main advantages of using Ni-Ti alloys are high shape memory effect

Table 6.1 Shape-memory alloys and their properties.

Properties	NiTi	Cu-Zn-Al	Cu-Al-Ni
Density (g/cm³)	6.4–8.5	7.8–8.0	7.1–7.2
Electrical conductivity	1–15	8–13	7–9
Tensile strength (N/mm²)	800–1000	400–700	700–800
Elongation (%)	40–60	10–15	5–6
Maximum Ag-Temperature (°C)	120	120	170
Maximum One-Way Effect $\varepsilon_{1\,max}$ (%)	8	4	5
Maximum Two-Way Effect $\varepsilon_{2\,max}$ (%)	6	2	2
Overheatable (Short time) up to (°C)	400	180	300

(SME), high stability of the effect, good corrosion resistance good over-heat ability, and the high volume of work [23]. Even though copper-based alloys Cu-Zn-Al have smaller SMEs, overheat ability, and corrosion resistance than that of NiTi, they are less expensive and also less stable. But interesting fact is its enhanced electric and thermal conductivity [16]. Unfortunately, these copper-based allows possessing ductile nature and lesser stability when compared with conventional materials [22]. Table 6.1 shows the physical, mechanical, and some shape memory properties; the three above-mentioned alloys were given as ranges or maximum values.

6.2.1.4   Factors Affecting SMA Actuators

6.2.1.4.1   Mechanical Stress
From redesigned Clausius-Clapeyron Equation, the change or effect of the switching temperatures with the application of extrinsic stress can be easily calculated:

$$V. \ (d\sigma/dT) = -(\Delta H_{M \longrightarrow A}) / (T_0{}^*{}_\varepsilon)$$

In order to achieve phase transformation, there is a stress ($d\sigma$) rise on the actuator material that resulted in rise in temperature (dT) than that of standard temperature T_0. Their stress effect occurred as a result of these in mechanical system, disturbs the system stability, and promotes actuation, which is more practically applicable for Nitinol wire actuators. In some cases, as all transformation temperatures, the external stresses may itself adjust the performance of the system to the specified environmental temperature necessities.

6.2.1.4.2   Fatigue
Even though SMA's are smart materials, to use it as an actuator, it needs to show structural properties to meet some mechanical or structural requirements. When we apply high stress to the thin wire actuator materials, there may occur fatigue and fractures. Thus, we have to study these structural fatigue properties linked to fatigue mechanisms and it will decrease lifetime of actuator [24]. Among these, the functional fatigues are mainly depended on the difficulty of the transformation of martensitic phase on the microstructure. There are thermal and mechanical cycling effects observed on the shape memory actuators which determine the functional properties of the materials. Wagner *et al.* analyze the possible combinations of the mechanical and functional loading. In the first stages of the

cycling process, thermal or thermo-mechanical–related functional properties occur, and later, it stabilizes. But the chances for structural fatigue persists at any time of the working process due to the acting mean stress and stress amplitude.

6.2.1.5   Applications of SMA in Actuation

Based on these characteristics, SMMs are used in many applications such as:

- Coupling mechanisms in tubes
- Ventilation techniques
- Thermal protection
- Heat engines
- Fire protection techniques
- Automotive techniques
- Solar techniques
- Actuator for robotics
- Actuator for electronics

6.2.2   Shape Memory Polymer Actuators

Actuators are nano-, micro-, and macroscale working mechanical devices that can change their shape with respect to the environmental conditions [25–27]. Actuators are widely implemented in microelectronics, biomedical field, micro-fabrication, medical field, lab-on-a-chip systems, and embedded systems [28]. The design and development of actuator design are very complex since it includes material as well as engineering aspects. There are different actuator types based on the materials used, such as metals and their oxides, polymerized materials, and hydrogels [25, 29]. Between these types, polymer-based receives much attention because of their attractive properties [26, 30–33]. Polymers may differ in their chemical and physical structure as soft/viscoelastic or hard/glassy state, which makes the formation of delicate smooth actuators for manipulating biological existing body molecules or cells and tough actuators for manipulating metallic elements and alloys [34]. Some polymers may be sensitive to pH, temperature, light, lasers, or bio-signals, and design of actuators with them enables the stimulus as the sensitive part [35]. Also, actuators from biodegradable and biocompatible polymers find application in existing biological systems and they adhere there.

Even though, utilizing SME in real engineering applications was started in 1906s, the usage of heat-shrink polymeric materials was widely used earlier in the form of protecting wire/cable bundles [6]. Now, there is a large growth that occurred in SMM network from different metallic alloys to polymers, which include gels and liquid crystals to ceramics, hybrids, etc. [5, 8, 9]. Depending on the materials, the stimuli used also varied such as heat-induced, photo-induced, chemo-induced, and induced using magnetic fields [5, 10, 36]. Most often, the term SME is confused with SCE (Shape Changing Effect), which is another shape triggering mechanism. The major difference among the two was that SME can trigger the shape recovery process on demand with the application and SCE produces a sudden response with the exposure of the correct triggering. Mechano-responsive portions that belong to SCE usually possess the loading/unloading tendency within its elastic range and revertible expanding and contractive tendency visible seen in hydrogels [13]. The energy barrier level difference between SCE and SME is discussed and shown in Figure 6.5.

From Figure 6.5, H indicates the energy barrier, and its increased value indicates a proper triggering method. If the value of H' is low, SMM has the ability to switch between its permanent and temporary shapes instantly. In the 1990s, the SMM community became more and more relevant due its major applications in the field of SMT [13]. This SME paved/inspired the scientific world, which resulted in the emergence of novel applications using conventional materials/technologies and also reshaped the design of product in many ways [11, 12, 14, 37–41]. We can integrate more and more functions such as sensing, actuation, etc., into the SMM and it behaves as "The material is the machine" [42] which is ready for implementation

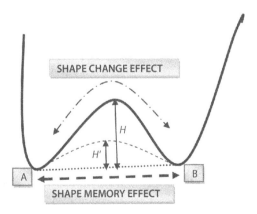

Figure 6.5 Shape memory effect (SME) and shape change effect (SCE).

purpose. Research studies show that almost all polymeric materials belong to chemo-responsive or thermo-responsive SMM type [9]. To design an efficient SME in a material, we have to design some mechanisms to set its temporary shape for shape recovery. The major fixation mechanisms possible [23] are described below:

- a. Dual-state mechanism (DSM)
- b. Dual-component mechanism (DCM)
- c. Partial transition mechanism (PTM)

a. Dual-state mechanism (DSM)

Some SMMs may be of one state/phase which depends on the environmental condition(s). Figure 6.6 explains the DSM mechanism scheme. The scheme (a) represents a thermo-responsive alloy that works on temporary martensite stage at lower temperature and austenite stage at higher temperature values. (b) Shows a magneto-responsive SMA, in which a magnetic field is required for deformation. The addition of elastomer materials such as silicone into the SMM favors the shape recovery by the means of glass transition temperature (T_g). The shape recovery occurs on the basis of glass transition and melting temperature is shown in diagram under (c).

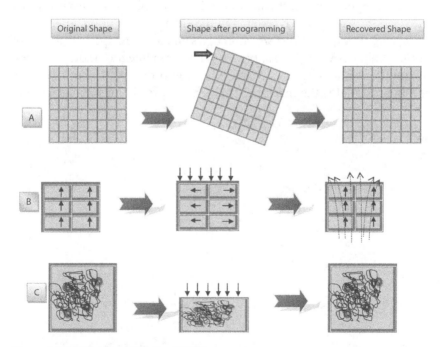

Figure 6.6 Dual-state mechanism (DSM).

b. Dual-component mechanism (DCM)

Some SMMs itself possess a dual-phase or dual-segment due to their internal segment parts, which differs it from DSM (Figure 6.7). Let's consider an example: polyurethane contains two parts. A soft segment (helps in transition) developed from polyol groups used inside during synthesis and a hard segment (which is elastic in nature) developed inside die to the reaction of diisocyante in their molecular structure. These two segments cause phase separation inside the structure. One example of the dual-phase system is robbery material integrated with wax segments, which possess excellent thermo-responsive SME [43].

c. Partial transition mechanism (PTM)

PTM is a mechanism occurred due to the combinational effect of DSM and DCM (Figure 6.8). Polymeric material's cross-linking property and some rubbery segments have a tendency to hold some power within it after its design. This power stored in it is one of the most sufficient triggering methods to enhance the final shape recovery. In this PTM, a small part of the SMM is soft as well as flexible type and the other majority portion, which is hard enough and takes the role of the strong elastomer matrix. Making use of this working mechanism, SME can be enabled in melting glue and wax by taking their meting point a concern.

In the majority of real-time practices, more than one or a simultaneous combination of these mechanisms may occur. The combined effects

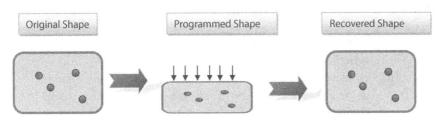

Figure 6.7 Dual-component mechanisms (DCM).

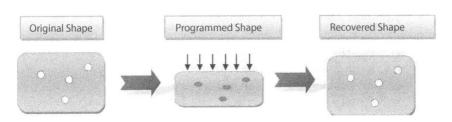

Figure 6.8 Partial-transition mechanism (PTM).

of three design mechanisms are widely adopted in polymeric materials with temperature or thermo-responsive effect (TME) [15, 44]. These DSM or DCM and PTM provide the necessarily required environment for the SME. Another effective way to measure SME performances is to use the terms shape fixity ratio (R_f) and shape recovery ratio (R_r). Shape fixity ratio (R_f) is the measure of ability of material to fix the shape transformation and shape recovery ratio (R_r) is ability of a material to recover its original shape. Both are calculated from the final recovered material after the exposure of right stimulus to it. These two ratios are not only a material-dependent so that we can tailor by varying compositions. But these are affected by programming parameters such as stress-strain, applying temperature, etc. [45]. There are also many reported materials that can be triggered using multiple stimuli. Some polyurethane materials with SME can be stimulated using moisture or ethanol [46, 47]. Also, some SMM may have both the SME and SCE properties. A hydrogel can be considered as the best example, which is in SMM stage when moisture content is and becomes viscoelastic as the water content increases, which belongs to SCE type. SME in these types of polymeric materials is not limited to gels, liquid crystal materials, etc. [3, 9, 48–52], but also proteins (in hair, nail, silk, etc.) [9].

6.2.2.1   Classification of Polymeric Actuators

Polymeric actuators have many novel features and some of them are listed below:

(i)   The actuation of these materials may not affect atmospheric wet and dry conditions.
(ii)   Glass transition or the melting temperatures, which helps in the shape deformation, can be designed according to our choice by changing the polymers and their mixing ratios.
(iii) Most of the materials are eco-friendly green and biologically compatible so that it is used in medical field also.

Polymeric actuators can be classified and discussed below:

a.   Actuators based on elastic relaxation
b.   Liquid-crystalline (LC) actuators
c.   Revertible volume change actuators
d.   Surface tension actuators

a. Actuators based on elastic relaxation

Consider dielectric elastomeric actuators as an example of this type. This class of actuators has the arrangement in such a way that some elastomers which act as a dielectric material are placed between the two electrodes [53]. Here, applied voltage acts as the stimulus which helps actuator to generate some energy between two electrodes, and hence, it compresses the rubbery material fixed between them. Then, the applied voltage is removed, the force decreases, and electrodes move to their original position and dielectric material relaxes. If the actuator materials possess an inhomogeneous cross-linking density, it also results in the elastic shape recovery [54]. Another example of these relaxation-type shape memory polymers

b. Liquid-crystalline (LC) actuators

These classes of actuators mainly depend on mesogen, which exists in mesophase and known as the fundamental units of liquid crystals. The orientation of these mesophase groups undergoes change in their alignment due to the re-arrangement in their order and this is the working principle of LC [55]. Generally, these actuators are revertible stimuli-responsive types and can be used for working in a solvent-free environment [56]. This allignment of mesogen groups can be controlled by means of technique called photopolymerization. Thus, obtained LC possesses a defined microstructure and can be used as the hierarchical materials. In polymer integrated liquid-crystals, the properties can be enhanced by either stretch or shrink anisotropically along the direction of the orientation of the order parameter changes.

c. Revertible volume change actuators

Hydrogels, which can swell reversible in water, are the one example of this type of actuators [26]. Here, temperature-sensitive polymers with significantly low thermal expansion coefficient will not permit transformation. During the first-order transition phase (melting/crystallization), the change in volume reaches its higher level, and thus, it results in huge transformation of shape. These actuator series belong to revertible type.

d. Surface-tension actuators

Surface tension means the force experienced on the surface layer of fluid. These types of actuators are small with the major driving force that is enough surface tension and belongs to the irreversible type. Crystalline polymers which have moderate melting point are used as the actuation materials [57].

Applications of Polymer Actuators

Some of the major applications are described below:

1. Sensors
If the material will undergo some macroscopic changes in their shape, with the application of stimulus can be used for the design of sensors or can be used as a sensor material [58, 59]. There were broad examples of hydrogels that can be which is already reported for sensing applications in various science and engineering applications. In AFM cantilevers, one side of the actuator material is coated with a stimuli-responsive hydrogel so that it will undergo rapid deform due to the expansion stage in hydrogel [59, 60]. The laser light detects the occurred deformation of cantilever.

2. Imaging Technologies
Hydrogel actuators are employed in the manufacture of different lenses having variable focal lengths because of its transparent nature. Another example is stimuli-responsive hydrogel, which is placed on a ring with an isolating layer to separate the water content and viscous oil [61, 62]. By activating this hydrogel with appropriate trigger, there is simultaneous inflation and dilation inside the gel. Thus, there occur some variation in water content, and as a result of this, there are net volume changes that cause some force change at the isolating boundary which has a major role in affecting the design of imaging tools.

3. Controlling Liquid Flow
This is considered as the one among the various potential applications of polymeric actuators [63–67]. Consider pH-sensitive hydrogel pieces that can behave like a smart valve. The valve has the ability to activate and deactivate in response to solution pH [68].

4. Walkers and Swimmers
By cyclical stimuli, actuator shape change can be transformed in walking or swimming [69]. Polyelectrolyte hydrogel is subjected to cyclical stimuli's that will make them have the ability to act in swim mode, as a result of their motion due to their shape transformation [70, 71].

5. Three-Dimensional Microfabrication
6. Smart Textiles
7. Surgery
8. Switchable Surfaces

References

1. Huang, W., Ding, M.Z., Wang, C.C., Wei, J., Zhao, Y., Purnawali, H., Shape memory materials. *Mater. Today*, 13, 7–8, 54–61, 2010, https://doi.org/10.1016/S1369-7021(10)70128-0.
2. Otsuka, K. and Wayman, C., Mechanism of shape memory effect and superelasticity, in: *Shape Memory Materials*, Cambridge University Press, pp. 27–48, 1998.
3. Huang, W., Yang, B., Zhao, Y., Ding, Z., Thermo-moisture responsive polyurethane shape-memory polymer and composites: A review. *J. Mater. Chem.*, 20, 3367–3381, 2010.
4. Funakubo, H. and Kennedy, J., *Shape memory alloys*, p. xii+ 275, 15 x 22 cm, Illustrated, Gordon and Breach Publishers, 1987.
5. Sun, L., Huang, W.M., Ding, Z., Zhao, Y., Wang, C.C., Purnawali, H., Tang, C., Stimulus-responsive shape memory materials: A review. *Mater. Des.*, 33, 577–640, 2012, https://doi.org/10.1016/j.matdes.2011.04.065.
6. Gunes, I.S. and Jana, S.C., Shape memory polymers and their nanocomposites: A review of science and technology of new multifunctional materials. *J. Nanosci. Nanotechnol.*, 8, 1616–1637, 2008.
7. Huang, W., On the selection of shape memory alloys for actuators. *Mater. Des.*, 23, 11–19, 2002.
8. Behl, M. and Lendlein, A., Actively moving polymers. *Soft Matter*, 3, 58–67, 2007.
9. Ahir, S.V., Tajbakhsh, A.R., Terentjev, E.M., Self-assembled shape-memory fibers of triblock liquid-crystal polymers. *Adv. Funct. Mater.*, 16, 556–560, 2006.
10. Lendlein, A., *Shape-memory polymers*, Springer, 2010.
11. Abrahamsson, P. and Bjämemo, R., *Ecomaterials*, pp. 1171–1174, Elsevier BV, 1994.
12. Farzin-Nia, F. and Inst Mech, E., Orthodontic applications of shape-memory technology, in: *Medical applications for shapememory alloys*, pp. 41–50, Professional Engineering Publishers, London, 1999.
13. Sun, L. and Huang, W., Wet to shrink: An approach to realize negative expansion upon wetting. *Adv. Compos. Mater.*, 18, 95–103, 2009.
14. Zhao, Y., Huang, W.M., Wang, C.C., Thermo/chemo-responsive shape memory effect in polymers: A sketch of working mechanisms, fundamentals and optimization. *J. Polym. Res.*, 19, 9952, 2012, https://doi.org/10.1166/nnl.2012.1412.
15. Bhattacharya, K. and James, R.D., The material is the machine. *Science*, 307, 53–54, 2005.
16. Tautzenberger, P., *European Symposium on Martensitic Transformations*, pp. 213–222, EDP Sciences, 1989.
17. Tautzenberger, S. and Stöckel, D., *Proc. 30th Holm Conf. on Electrical Contacts*, pp. 449–454, 1984.
18. Tautzenberger, P. and Stöckel, D., Gedächtnis-Effekt und technisch anwendbare Legierungen. *Z. Wirtsch. Fertigung*, 78, 486–488, 1983.

19. Bellouard, Y., Shape memory alloys for microsystems: A review from a material research perspective. *Mater. Sci. Eng.: A*, 481, 582–589, 2008.
20. Winzek, B., Schmitz, S., Rumpf, H., Serzl, T., Hassdorf, R., Tienhaus, S., Feydt, J., Moske, M., Quandt, E., *Mater. Sci. Eng. A*, Taylor and Francis group, 378, 2004, 40–46.
21. Perkins, J. and Sponholz, R., Stress-induced martensitic transformation cycling and two-way shape memory training in Cu-Zn-Al alloys. *Metall. Trans. A*, 15, 313–321, 1984.
22. Duerig, T.W., Albrecht, J., Gessinger, G.H., A shape-memory alloy for high-temperature applications. *JOM*, 34, 14–20, 1982.
23. Buehler, W., Gilfrich, J., Wiley, R., Effect of characteristic temperatures of thermoelastic martensitic properties of alloys near composition TiNi. *J. Appl. Phys.*, 34, 1475–1476, 1963.
24. Mertmann, M. and Vergani, G., Design and application of shape memory actuators. *Eur. Phys. J. Spec. Top.*, 158, 221–230, 2008.
25. Randhawa, J.S., Laflin, K.E., Seelam, N., Gracias, D.H., Chemically Controlled Miniature Devices: Microchemomechanical Systems. *Adv. Funct. Mater.*, 21, 13, 2011.
26. Ionov, L., Biomimetic hydrogel-based actuating systems. *Adv. Funct. Mater.*, 23, 4555–4570, 2013.
27. Geryak, R. and Tsukruk, V.V., Reconfigurable and actuating structures from soft materials. *Soft Matter*, 10, 1246–1263, 2014.
28. Kim, S., Laschi, C., Trimmer, B., Soft robotics: A bioinspired evolution in robotics. *Trends Biotechnol.*, 31, 287–294, 2013.
29. Mirfakhrai, T., Madden, J.D., Baughman, R.H., Polymer artificial muscles. *Mater. Today*, 10, 30–38, 2007.
30. Liu, Z. and Calvert, P., Multilayer hydrogels as muscle-like actuators. *Adv. Mater.*, 12, 288–291, 2000.
31. Jager, E.W., Smela, E., Inganäs, O., Microfabricating conjugated polymer actuators. *Science*, 290, 1540–1545, 2000.
32. Osada, Y., Okuzaki, H., Hori, H., A polymer gel with electrically driven motility. *Nature*, 355, 242, 1992.
33. Wen, H., Zhang, W., Weng, Y., Hu, Z., Photomechanical bending of linear azobenzene polymer. *RSC Adv.*, 4, 11776–11781, 2014.
34. Zhang, Y. and Ionov, L., Actuating porous polyimide films. *ACS Appl. Mater. Interfaces*, 6, 10072–10077, 2014.
35. Stuart, M.A.C., Huck, W.T.S., Genzer, J., Müller, M., Ober, C., Stamm, M., Sukhorukov, G.B. *et al.*, Emerging applications of stimuli-responsive polymer materials. *Nat. Mater.*, 9, 101, 2010, doi: 10.1038/nmat2614.
36. Otsuka, K., Cambridge University Press, 1998.
37. Toensmeier, P.A., Shape memory polymers reshape product design. *Plast. Eng.*, 61, 10–11, 2005.

38. Sun, L., Huang, W.M., Wang, C.C., Ding, Z., Zhao, Y., Tang, C., Gao, X.Y., Polymeric shape memory materials and actuators. *Liq. Cryst.*, 41, 277–289, 2014, https://doi.org/10.1080/02678292.2013.805832.

39. Wischke, C., Neffe, A.T., Steuer, S., Lendlein, A., Evaluation of a degradable shape-memory polymer network as matrix for controlled drug release. *J. Controlled Release*, 138, 243–250, 2009.

40. Wischke, C. and Lendlein, A., Shape-memory polymers as drug carriers—A multifunctional system. *Pharm. Res.*, 27, 527–529, 2010.

41. Feng, Y., Zhang, S., Wang, H., Zhao, H., Lu, J., Guo, J., Behl, M., Lendlein, A., Biodegradable polyesterurethanes with shape-memory properties for dexamethasone and aspirin controlled release. *J. Controlled Release*, 152, 2011, doi: 10.1016/j.jconrel.2011.08.098.

42. Kratz, K., Voigt, U., Lendlein, A., Temperature-Memory Effect of Copolyesterurethanes and their Application Potential in Minimally Invasive Medical Technologies. *Adv. Funct. Mater.*, 22, 3057–3065, 2012.

43. Sun, L. and Huang, W., Thermo/moisture responsive shape-memory polymer for possible surgery/operation inside living cells in future. *Mater. Des. (1980-2015)*, 31, 2684–2689, 2010.

44. Huang, W.M., Yang, B., Fu, Y.Q., *Polyurethane shape memory polymers*, CRC Press, 2011.

45. Huang, W.M., Song, C.L., Fu, Y.Q., Wang, C.C., Zhao, Y., Purnawali, H., Lu, H.B., Tang, C., Ding, Z., Zhang, J.L., Shaping tissue with shape memory materials. *Adv. Drug Delivery Rev.*, 65, 515–535, 2013, https://doi.org/10.1016/j.addr.2012.06.004.

46. Xie, T., Tunable polymer multi-shape memory effect. *Nature*, 464, 267, 2010.

47. Sun, L. and Huang, W.M., Mechanisms of the multi-shape memory effect and temperature memory effect in shape memory polymers. *Soft Matter*, 6, 4403–4406, 2010.

48. Sun, L., Huang, W.M., Wang, C.C., Zhao, Y., Ding, Z., Purnawali H., Optimization of the shape memory effect in shape memory polymers. *J. Polym. Sci., Part A: Polym. Chem.*, 49, 3574–3581, 2011, https://doi.org/10.1002/pola.24794.

49. Wang, C.C., Zhao, Y., Purnawali, H., Huang, W.M., Sun, L., Chemically induced morphing in polyurethane shape memory polymer micro fibers/springs. *React. Funct. Polym.*, 72, 757–764, 2012.

50. Osada, Y. and Gong, J.P., Soft and wet materials: Polymer gels. *Adv. Mater.*, 10, 827–837, 1998.

51. Iqbal, D. and Samiullah, M., Photo-responsive shape-memory and shape-changing liquid-crystal polymer networks. *Materials*, 6, 116–142, 2013.

52. Ilnytskyi, J.M., Saphiannikova, M., Neher, D., Allen, M.P., Modelling elasticity and memory effects in liquid crystalline elastomers by molecular dynamics simulations. *Soft Matter*, 8, 11123–11134, 2012.

53. Bozlar, M., Punckt, C., Korkut, S., Zhu, J., Foo, C.C., Suo, Z., Aksay, I.A., Dielectric elastomer actuators with elastomeric electrodes. *Appl. Phys. Lett.*, 101, 091907, 2012, https://doi.org/10.1063/1.4748114.

54. Jamal, M., Zarafshar, A.M., Gracias, D.H., Differentially photo-crosslinked polymers enable self-assembling microfluidics. *Nat. Commun.*, 2, 527, 2011.
55. Yu, Y., Nakano, M., Ikeda, T., Photomechanics: Directed bending of a polymer film by light. *Nature*, 425, 145, 2003.
56. Ohm, C., Brehmer, M., Zentel, R., Liquid crystalline elastomers as actuators and sensors. *Adv. Mater.*, 22, 3366–3387, 2010.
57. Azam, A., Laflin, K.E., Jamal, M., Fernandes, R., Gracias, D.H., Self-folding micropatterned polymeric containers. *Biomed. Microdevices*, 13, 51–58, 2011.
58. Singamaneni, S., LeMieux, M.C., Lang, H.P., Gerber, C., Lam, Y., Zauscher, S., Datskos, P.G., *et al.*, Bimaterial microcantilevers as a hybrid sensing platform. *Adv. Mater.*, 20, 653–680, 2008, https://doi.org/10.1002/adma.200701667.
59. Bashir, R., Hilt, J., Elibol, O., Gupta, A., Peppas, N., Micromechanical cantilever as an ultrasensitive pH microsensor. *Appl. Phys. Lett.*, 81, 3091–3093, 2002.
60. Hilt, J.Z., Gupta, A.K., Bashir, R., Peppas, N.A., Ultrasensitive biomems sensors based on microcantilevers patterned with environmentally responsive hydrogels. *Biomed. Microdevices*, 5, 177–184, 2003.
61. Dong, L., Agarwal, A.K., Beebe, D.J., Jiang, H., Adaptive liquid microlenses activated by stimuli-responsive hydrogels. *Nature*, 442, 551, 2006.
62. Richter, A. and Paschew, G., Optoelectrothermic Control of Highly Integrated Polymer-Based MEMS Applied in an Artificial Skin. *Adv. Mater.*, 21, 979–983, 2009.
63. Yu, Q., Bauer, J.M., Moore, J.S., Beebe, D.J., Responsive biomimetic hydrogel valve for microfluidics. *Appl. Phys. Lett.*, 78, 2589–2591, 2001.
64. Arndt, K.F., Kuckling, D., Richter, A., Application of sensitive hydrogels in flow control. *Polym. Adv. Technol.*, 11, 496–505, 2000.
65. Dong, L. and Jiang, H., Autonomous microfluidics with stimuli-responsive hydrogels. *Soft Matter*, 3, 1223–1230, 2007.
66. Beebe, D.J., Moore, J.S., Yu, Q., Liu, R.H., Kraft, M.L., Jo, B.-H., Devadoss, C., Microfluidic tectonics: A comprehensive construction platform for microfluidic systems. *Proc. Natl. Acad. Sci.*, 97, 13488–13493, 2000, https://doi.org/10.1073/pnas.250273097.
67. Eddington, D.T. and Beebe, D.J., Flow control with hydrogels. *Adv. Drug Delivery Rev.*, 56, 199–210, 2004.
68. Smela, E., Conjugated polymer actuators for biomedical applications. *Adv. Mater.*, 15, 481–494, 2003.
69. Ma, Y., Zhang, Y., Wu, B., Sun, W., Li, Z., Sun, J., Polyelectrolyte multilayer films for building energetic walking devices. *Angew. Chem. Int. Ed.*, 50, 6254–6257, 2011, https://doi.org/10.1002/anie.201101054.
70. Morales, D., Palleau, E., Dickey, M.D., Velev, O.D., Electro-actuated hydrogel walkers with dual responsive legs. *Soft Matter*, 10, 1337–1348, 2014.
71. Kwon, G.H., Park, J.Y., Kim, J.Y., Frisk, M.L., Beebe, D.J., Lee, S.H., Biomimetic soft multifunctional miniature aquabots. *Small*, 4, 2148–2153, 2008, https://doi.org/10.1002/smll.200800315.

7

Stimuli-Responsive Conducting Polymer Composites: Recent Progress and Future Prospects

Ajahar Khan* and Khalid A. Alamry†

Faculty of Science, Department of Chemistry, King Abdulaziz University, Jeddah, Saudi Arabia

Abstract

Stimuli-responsive electroactive polymers-based conducting polymers (CPs) composites have been comprehensively studied because of their adaptable properties, simple processability, low cost, robust behavior, and functional properties in respond to various external stimuli. Therefore, the CPs composites have grown to be the material of preference for a variety of grown-up and forefront technologies. The last few decades have witnessed a quick expansion of development and design of such CPs composites, which originate tremendous opportunities for examining their applications in promising fields such as soft robotics, biomedical applications, artificial muscles, and electromechanical engineering. To guide the future development, there is a need to understand the progress of the respective materials with their properties and applications in different fields. This chapter reports the current progress of stimuli-responding CPs composites to understand the mechanical behavior of these materials with respect to electrical-, photo-, and thermo-responsive stimuli. An overview of most widely used CPs (such as polypyrrole and polianiline)–based composite materials with respect to their uses and operational mechanisms, as well as a specified account of CPs as next generation actuators are also described.

Keywords: Conductive polymers, bending actuators, artificial muscles, robotics

Corresponding author: arkhan.029@gmail.com
†*Corresponding author*: kaalamri@kau.edu.sa

Inamuddin, Rajender Boddula and Abdullah M. Asiri (eds.) Actuators: Fundamentals, Principles, Materials and Applications, (159–186) © 2020 Scrivener Publishing LLC

7.1   Introduction

Stimuli-responsive electroactive polymers (EAPs) symbolize a vast variety of composite materials that can provide substantial deformation in their structures and can modify their properties in respond to different external stimuli. Stimuli-responding EAPs can be classified on the basis of their external stimuli, such as magnetic-, electrical-, thermo-, photo-, humidity-, pressure-, and pH-responsive composite materials, etc. [1–10]. Among the EAPs such as ionic gels, dielectric elastomers, ionic polymer metal composites (IPMCs), and carbon nanotubes-based actuators, conducting polymers (CP) composites–based bending actuators are studied to be attractive because they are operated electrically at low applied voltage and give response to different physical and chemical stimuli [11–14]. Majority of the CP composites studied for soft robotics, electronics, and electromechanical robotic engineering applications. Therefore, the research interests are growing towards photo- [15], humidity- [16–18] pressure- [19–21] electrical- [22, 23], and pH-responsive [24–28] composites due to their possible applications in the fields of environmental sensing, micro robotics, and biomedical systems. Particularly, ionic polymer soft actuators-based on CPs composites have important and notable properties as well as the capability to display large bending movement at low applied voltages (<5 V), low noise, space-saving biomimetic application, and less power utilization. Moreover, these types of ionic polymer soft actuators have been previously used as soft aquatic mini-robots [29–38]. Ionic polymer actuators based on CPs composites have relatively more strain and stresses, usually perform in an electrolyte solution as well as in open air [39–41].

The attention in developing actuating micro-devices has been escalating in the past decade as a consequence of soaring demand of biomimetic multifaceted mechanisms such as implantable neuronal devices, animal-like robots, tissue substitutes, etc. [42–44]. In this framework, CPs composites are proper candidates for such applications in view of their low actuation potentials, high mechanical properties and biocompatibility [45, 46]. The doping/un-doping processes can be used to attribute the electro-activity of CPs, when they are getting in touch with an ions source. The conformational changes of the entire structure is generated due to the contraction/expansion, when these ions are start moving outside and inside within the polymer chains [47]. In particular, polypyrrole (PPy) and polyaniline (PANI) are most widely used for fabricating actuators due to the prospect of obtaining different morphologies and shapes, which can enhance the attributes of the concluding device. PANI and PPy are very promising for commercial applications due to their higher conductivity and facile synthesis, good environmental strength, than many other CPs. The literature survey has shown that the deposition of PPy on mechanically immobile metal/flexible plastic membrane bilayers gives mechanical stable actuator devices [48, 49]. Large power density (150 W/kg), moderate active strains (2%), large dynamic stresses (10–30 MPa), and easy operation at low electric consumption (1–3 V) make PPy a striking platform for designing electroactive polymeric materials [50]. It has been already reported that the devices developed using PANI-based micro-fibers or tubes and porous films display enhanced functional properties than those based on classical films [51–53]. Therefore, the type of material and its morphology directly or indirectly influences the actuation performance by improving the surface to volume ratio. As the active surface is larger, the more ions will start moving inside/outside within the polymer structure and generate a macroscopic deformation. The methods carried out for fabricating artificial muscles were considering only the adhesive tapes with a CP film, coverage of thin metal layer, and the chemical synthesis of porous films. Gu *et al.* studied that the more specific area of the fabricated hybrid polyurethane/PANI parallel nanofibers enhances the incorporation of ions leading to large bending deformation. Moreover, the bundles form by the parallel nanofibers make the configuration which is very similar to the structure of natural muscular fibers [54]. This study shows that, CPs composites-based soft actuator attains extensively bigger macroscopic electroactive feedback for a given set of driving conditions.

7.2 Conductive Polymers (CPs)

First of all, Inganas and co-workers [55, 56] have given the information that CPs exhibit conjugated backbone of organic compound. They sandwiched an electrolyte solution between two layers of CP electrodes that could behave like actuators. Two most common types of CPs that widely used are PPy and PANi, apart from these other candidates acknowledged to date consist of poly(p-phenylene vinylene), trans-polyacetylene, and polythiophene [50, 57, 58]. The actuation mechanism in the CPs composite-based actuators produces on the base of reversible counter ion uptake and discharge along with inner solvent that occurs during redox cycling [50, 57–59]. The changes in oxidation state during electrochemically induction endorse analogous charge flux within the backbone of polymer chain, which then initiate notable fluctuation of ions to equilibrate the charge. These responsive ions cause the CPs composite structure to expand during oxidation or compact if it is reduced subsequently, thereby exhibiting convenient and cyclic bending deformation [44]. This electrochemical reactions requires low voltages (1–5 V) to commence this bending response and resulting actuation strains vary from 1 to ca. 40% [59, 60]. A capable feature of CPs composite-based actuator films is that their attainable force density, during calculated up to ~100 MPa is likely to be as more as 450 MPa [44]. However, CPs composites-based actuators endure from numerous considerable drawbacks such as electromechanical coupling (<1%) and low capability (on the order of 1%), [50] in adding to reasonably small precise actuation-strain plain (<12%) [61]. The diffusion method requisite for the ions to be absorbed in and excluded from the polymer and the internal resistance among the polymer backbone and electrolyte solution [50] slow down the actuation-strain rates [57]. Besides all these shortcomings, CPs composites-based soft actuators symbolize feasible candidates for modern technologies adopting low potential consumption and high energy density and reproducibly that can be used for emerging fields of engineering and medicine.

7.3   Consequences of CPs

The use of CPs as electrode material increases the opportunity of incorporating more functional groups in the area of electrode surface because of the ramified polymer network, tight adherence to electrode constituents, and capacity to modify the properties of CPs films by altering the parameters at the time of electrochemical polymerization [62]. CPs such as PPy and PANI films are fully stable at feasible conditions [23, 63, 64]. The monomers of PPy and PANI are easily oxidized and provide good redox properties and environmental stability [65]. The CPs nanocomposites are able to respond in light, temperature, humidity, and various chemo- and physical response and attain brilliant bending deformation beneath the influence of various external stimuli [66]. The nanocomposites of conducting PPy and PANI nanomaterials were suppose to be possible layers for the charge transfer and surface protection in ionic polymer actuators. In the nanocomposite matrix, CPs are considered to be novel electroactive relay along with metal nanoparticles as compared to the nanocomposites constructed using insulating polymers. As a consequence, the development of nanocomposite materials based on CPs composites may offer appealing characteristics with superior features in chemical sensor and actuator-based robotics and biomimetic applications, since their physical properties often vary noticeably in response to changes of the chemical environment.

7.4   Synthesis or Polymerization of Most Widely Used CPs for Actuator Applications Such as PPy and PANI

7.4.1   Polymerization of Polypyrrole

First time PPy was synthesized in 1912 [67]. Polypyrrole polymerized by conservative chemical routes is not soluble in regular solvents due to tough inter-chain interactions [68]. The two most common ways used to synthesize or polymerize PPy by different factors are: (a) chemical initiation of pyrrole monomers using oxidative agents [67, 69], (b) electrochemical polymerization by anodic current [70], and (c) photo-induced synthesis [71], etc. All the methods used to initiate polymerization of PPy have selective applications, e.g., chemical oxidative polymerization might be efficiently needed if large amount of PPy is required for the propose of chromatography columns application [72] or for numerous other purposes. The oxidative chemical [69] or biochemical polymerization [73] techniques can be used to develop PPy nanoparticles of desired shapes with proscribed size ranging from micrometers to nanometers containing a variety of inclusions. Moreover, it is confident to unvaryingly achieve over oxidation of this polymer, by initiation of chemical oxidative polymerization, which might display preference to molecules ranging from the small [74–76] to high molecular weight [77]. Moreover, the chemically oxidative polymerization is primarily created in the bulk amount of solution, where a few proportion of synthesized PPy is covering the surface of imported materials. It shows that this route is not proficient in compare to the deposition of PPy over some material surfaces. Photo-induced polymerization of PPy is striking in photolithographic application, because it permits change in synthesized morphology of PPy by altering the wavelength of exciting light [78], that might be useful for the blueprint of electronic chips. However, it is still not very often use for the polymerization of PPy due to sluggish light-induced polymerization.

Moreover, Ppy is approximately insoluble in common solvents, but doping with proper gents increase the solubility of PPy [79]. This is clearly indicated that deposition of this polymer usually by solvent evaporation is possible from the solution having dissolved polymer in the form of colloidal particles prior to precipitation [73]. However, the main disadvantage of fabricating PPy-based actuator using deposition method is poor attachment to the surface as compare to the deposited film developed through electrochemical polymerization route. Finally, above discussed abnormalities can be eliminated by adopting the electrochemical polymerization. This method permits deposition of PPy over the surface of electrodes present within electrochemical cell. Therefore, this method is required when the deposition of thin PPy layers as electrodes material is requested. By using electrochemical polymerization, the morphology and the thickness of deposited PPy film might be controlled by defining the applied potential and passing current within electrochemical cell [80].

7.4.2   Synthesis of PANI

According to the literature review, different ways are available to produce PANI including electrochemical, chemical, enzymatic, photo-induced, and many more. Some conventional methods used to synthesize or polymerize aniline into PANI are discussed as follows. In the solution polymerization, aniline monomers have been polymerized into PANI in the presence of chloroform and electro-polymerized in acrylonitrile solvent [81, 82]. Generally, due to low solubility in the convenient solvents, the processability of PANI is found to be poor. Therefore, the PANI synthesized using solution polymerization method exhibit improved processability as it is formerly present in a solution [81]. In interfacial polymerization, the reaction is conducted in the edge of two immiscible solvents. Aniline has been polymerized by this method using a composition of two non-miscible solvents, water and chloroform with various acids used as dopants. The reaction has induced by oxidizing agent at R.T. or elevated temperature range in presence/absence of surfactant followed by centrifugation to isolate the final product [83, 84] A range of PANI composites with different shapes and size can be produced utilizing this method, through owing suitable reaction parameters and correct selection of reagents [83, 84]. When in an organic medium (benzene) with sodium amide, the p-dichlorobenzene is treated at 220°C up to 13 h, a metathesis reaction occurs which results into the formation of PANI [85]. This gives rise to another way for the production of PANI without the requirement of aniline monomers [86]. The illustration of the metathesis reaction for PANI production is shown in Scheme 7.1.

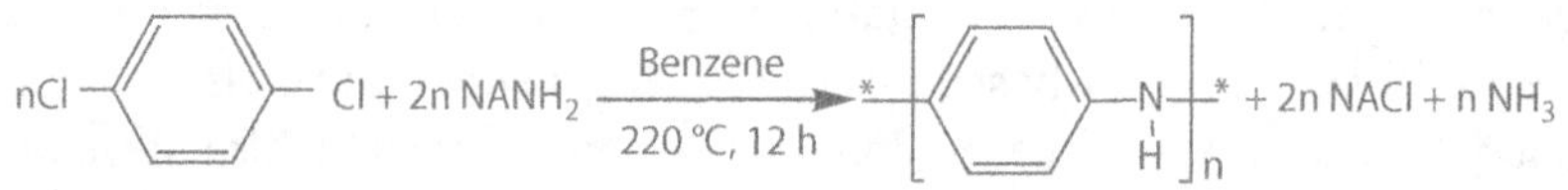

Scheme 7.1 The metathesis reaction for PANI production.

Yang *et al.* developed a molecularly massed PANI copolymer membrane using m-aminobenzene sulfonic acid and aniline through an oxidant ammonium peroxydisulfate at indium tin oxide substrate [86]. An ultra thin membrane can be obtain in a vapor phase by adopting self-assembling polymerization technique [87–89]. The polymerization of monomers of aniline in vapor phase directly produces a thin PANI film on polymeric membrane substrates (Figure 7.1). The 10 wt% solution consists of camphor sulfonic acid, FeCl3, and Fe (p-toluene sulfonate)$_3$ was made in methyl alcohol and layered over washed polymeric substrate membrane, e.g., polyimide, polyethylene terephthalate, polystyrene, polyvinyl chloride, etc., by spin or dip coating method followed by drying. Now, the dry film was treated with aniline vapor for 10–50 min in a sealed reaction assembly under diverse temperature range. Finally, the polymerized PANI films developed over the substrate were treated with methanol to eliminate unwanted byproducts and unused oxidant and dried for 4 min at temperature 80°C [106]. Using self-assembled method, Zhang *et al.* fabricated a PANI nanotube with 20–40 nm inner and 80–200 nm outer diameters [90]. The sonochemical method is started with the dropwise mixing of an acidic ammonium peroxydisulfate solution to aniline solution, in the same way as to that of conventional polymerization of PANI. However, the ultrasonic irradiation is responsible to accomplish polymerization. Using this technique, PANI nanofibers were synthesized with high polymer yields by Jing *et al.* [91, 92]. The main advantage of this method is its scalability in association with other methods such as rapid mixing reaction or interfacial polymerization.

The main attention to polymerize PANI goes towards electrochemical polymerization because it suggests a good route of polymerization with

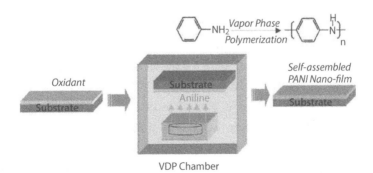

VDP Chamber

Figure 7.1 Schematic view to the procedure self-assembling PANI thin films through vapor-phase deposition polymerization [89].

excellent restriction over initiation and termination steps with good degree of technical potential. The PANI produced by electrochemical method polymerization is anticipated to be in pure form because it requires no other chemicals (oxidant and surfactant, etc.) as used in chemical route. Moreover, the utilization of fewer amounts of chemicals benefited for the healthy environment. This method is generally used to polymerize aniline monomers under galvanostatic current (constant), (b) potentiostatic potential (constant), and (c) under potential scans sweeping or cycling. The first route basically contains assembly of two electrodes immersed in a monomer containing electrolyte solution. In this assembly, a precise current is allowed to pass to develop PANI layer over the surface of platinum electrode. At constant potential polymerization, PANI powder is produced, which weakly remain on the electrode surface [93]. Moreover, by continuous cycling between predefined potentials, the electro-oxidation of aniline monomers produces an smooth PANI film, which tightly remains on the electrode [94, 95]. This obtained thin PANI film can be oxidized and reduced to optimize conductivity [96]. However, a thick PANI film can also be obtained and remove from the electrode in the form of free-standing film. This electro-chemical polymerization method can be used to produce fine nanowires of PANI. Usually, an inert electrode is utilized to carry out anodic oxidation of aniline. However, the common anode material would be conducting glass, platinum, metals such as Cu [97], Fe [98–100], graphite [96, 101–103], Au [104], stainless steel [105, 106], and vitreous carbon [107, 108] have been used. Gupta *et al.* studied the fabrication of PANI nanowires on a stainless steel electrode by electro-polymerization at 0.75 V vs. SCE [109].

The photo-induced synthesis comprises the photo-excitation of aniline monomers to attain PANI. The photons of UV-Vis region can be utilized to stimulate aniline monomers for polymerization in the transition metal salts solution [110]. This method generates a composite, where silver micro or nanowires are produced along with conducting PANI. The excitation wavelength defines the morphology of the produced PANI. A more fibrillar conformation is produce with visible light, whereas a more spherical morphology is obtained with UV light synthesis [78, 111].

7.4.3   CPs Composites as Actuators

CPs and their composites have been extensively explored for the use in soft actuators for micro robotics and artificial muscles applications [112]. Baker *et al.* reported that the CPs respond due to change in their structural configuration by the inclusion of solvating molecules and charged dopants or redox behavior by altering their oxidation state [113]. Single material bimorph soft actuator can even be made through PANI nanfibers by creating asymmetric membrane by using flash welding [113]. These PANI nafibers-based bimorph soft actuators respond fast to a large extent in compare to the other reported single or heterogeneous material-based bending actuators. The doping/dedoping of the membrane in acid and base is the stimulus for this actuator. This micro-actuator system may be used for the applications in a valve or sensor that operates by treating with acid or base. Gao *et al.* reported that composition of polyvinyl alcohol (PVA) with PANI nanofiber actuator demonstrates good response rate and tunable degree of actuation with high recoverability as compared to the actuator fabricated from pure PANI nanofiber [114]. Shedd *et al.* described the preparation of PANI nanfibers-based microstructures using electrophoretic deposition [115]. Here, an electrode batch charged the substrate to draw PANI-based nanofibers to the particular positions, while the neighboring electrodes were specified a reverse charge to avoid deposition at unwanted surface. After that, the obtained microstructures might be flash-welded to generate rigid material. The reported work projects its utilization for sensors application. However, if these flash-welded micro-membranes could be removed from the substrate, then the membranes would also be utilized as micro-actuator–based applications. There are numerous further CPs composite-based actuator structures that have been analyzed extensively with PANI, PPy, and their nanostructures. Zhang *et al.* determine the limitations of current density for PANI, to verify its probability in a large-strain ionic polymer-based electro-acoustic actuator. They concluded that in the conservative form, PANI doped with camphorsulfonic acid can hold a high current density up to 1,200 A cm^{-2} [116]. The electrostrictive copolymer operated with PANI was as efficient as gold under 1 Hz to 1 MHz frequencies in the temperature range of 50°C to 120°C. The current experimental work have led to the expansion of large breakdown field dielectric elastomer-based soft actuators by adding PANI nanoparticles as the large dielectric constant filler [117]. The dielectric elastomers behave like stimuli responding materials and demonstrate significant stress and strain under electric potential. In order to increase the actuation performance or bending strain, conducting fillers can be incorporated to amplify the dielectric constant of the elastomer without affecting the flexibility of the polymer mold [118]. The CPs (e.g., PANI and PPy) have more flexibility and good mechanical stability compared to metal fillers, when adopted in combination with valuable polymer composite actuator materials. Wang *et al.* reported improve electromechanical properties and dielectric constant with enhance electro-induced strain in a silicon rubber-based nanodielectric elastomer soft actuator composited with PANI nanorods. Kim *et al.* studied that the actuator fabricated using the blend of PANI nanorods and Nafion showing enhanced actuating performance compared to that made up of pristine Nafion [119]. Uh *et al.* reported that electrolyte free actuator-based PANI microfiber in the vicinity of magnetic field undergoes flapping wing motion, which shows the immense potential in micro-robotics, biomimetic, and artificial muscles field applications [120]. M. Beregoi *et al.* [121] design a particular architecture to fabricate lightweight, inexpensive, and easy to examine class of soft actuator films, consisting of Au and eggshell membrane (ESM) deposited with PPy. The PPy-coated Au/ESM (Au/ESM/PPy) membranes were analyzed in controlled humidity environment and ambient atmosphere, the recorded movements confirms their versatility. By varying the humidity, the PPy-based actuators deform, grasp, and liberate fragile and small weight objects (Figure 7.2). Liu *et al.* fabricated an ionic polymer-based actuator working in air by sandwiching PVA-H_2SO_4 electrolyte gel between reduced graphene oxide/PANI nanocomposite electrodes. Bending deformation

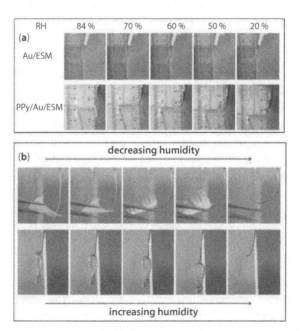

Figure 7.2 (a) The bending movement of the Au/ESM-based actuating membrane and the curling behavior of the Au/ESM/PPy-based actuator membrane due to reduction of the relative humidity. (b) The lifting of lightweight objects by the Au/ESM/PPy-based actuator due to altering relative humidity [121].

occurs due to migration of ions that produces differences in the volume between the cathode and anode under an applied electric field. This type of nanocomposite electrode because of the synergistic combination effects of PANI nanowires and GO sheets composition display brilliant electro-chemical stability over thousands of cycle [122]. This behavior is ascribed to the widespread sovereignty of expansion of the CPs composites-based soft actuators indifferent field of science, medical, and engineering.

7.4.4   CPs as Electrodes for Actuators

Qing Liu *et al.* proposed a new type of ionic polymer actuator working in air using poly(vinyl alcohol)/H_2SO_4 electrolyte gel sandwiching between reduced graphene oxide/PANI (r-GO/PANI) nanocomposite membrane electrode. In this actuator, the nanoparticles of PANI were evenly orna-mented through *in situ* polymerization over the r-GO film surfaces [122]. The combination of the high electrical conductivity, supramechanical properties, and large surface area of r-GO along with exceptional electro-chemical behavior of the conducting PANI consequences in the fabrication of composite material electrode attaining high electrochemical capacitance as well as good mechanical properties. Moreover, the experimental analysis reveals that the actuator was competent to stimulate at low applied poten-tial (0.5 V) exclusive of swapping actuator strokes and exhibit superb actu-ation durability. The actuation strain of the fabricated r-GO/PANI-based actuator possibly achieve to 0.327%, with respect to a 30 MPa stress at 0.5-V operating potential. This excellent actuation execution was supposed due to the synergistic combination of the PANI and r-GO component.

7.4.5   Tunable CP Membrane-Based Actuators

J. Gao *et al.* developed porous asymmetric PANI membranes through phase-inversion technique, and studied its bending deformation nature due to sorption/desorption of chemical vapors [123]. The asymmetrical configuration of the PANI film's cross section was responsible for bending-recovery movement. The intense side has more volume expansion sub-sequent to absorbing organic vapors than the other porous side, this bigger volume expansion responds to bending of the membrane toward the porous side. The CP-based membranes demonstrate bending-recovery behavior, when treated with organic vapors and retrieved after taking away from organic vapors due to the sorption/desorption. The organic vapors sorption causes extension of PANI backbone to give an extensive conformation, which increases the volume of the polymer membrane.

After organic vapors desorption, the PANI backbone regain to a compress configuration, cause the polymer membrane to return to its initial volume. This result obtained for porous asymmetric PANI membranes were analogous to that realized for electrochemically developed PPy-based film actuators, where films of PPy in the solid state go through fast and intensive bending-reinstatement deformation. Bending reinstatement deformation in the film of PPy was occurs generally due to the anisotropic and reversible aqueous vapor adsorption [124, 125].

7.4.6   Nanoporous Gold/CPs Composite-Based Metallic Muscles

Nanoporous metals–based metallic muscles experience some serious drawbacks due to the conventional use of electrolyte solution for actuation. Usage of electrolyte solution due to the depressed ionic conductivity stops the metallic muscles for working in waterless environmental conditions and eliminates larger actuation rate. This is the main cause to eliminate the metallic muscles for the integration into miniaturized systems [126]. In order to avoid all these restrictions, there is a need of developing an approach for metallic muscles, which is electrolyte-free [127]. Moreover, the following facts are prerequisite to breach within the area of artificial muscles applications: (i) quick actuation response, (ii) no use of aqueous or solid electrolyte, and (iii) solitary actuating assembly as found in piezoelectric-based systems. To meet this demand, CPs are feasible for developing an approach free from electrolyte to locate metallic muscles for work through metal/polymer edge that fulfill all the need requisite to avoid three above-mentioned hurdles. The micro or nano level coating of PANI doped with H_2SO_4 was developed onto the nanoporous gold surface. Sulfate anions from dopant co-adsorbed into the coating of PANI matrix were oppressed to adjust the nanoporous metal surface stress and consequently create noticeable dimensional adjustments in the metal. The strain rates attain for the single-constituent nanoporous metal/PANI composite material-based actuator are three times higher in amount in compare to the usual three-constituent nanoporous metal/electrolyte–based actuator.

7.4.7   CP-Based Nano-Fibrous Bundle for Artificial Muscles

Recently, a number of research groups have developed new methods intended to calculating the alignment and shape of nanofibers with customized electric field [128–130]. Spinks *et al.* demonstrated well-drawn PANI-based fibers composited with CNT to develop highly stable artificial muscles [131]. However, various research groups have been developed PANI films or microfibers-based actuators, but [113, 132] an proficient actuator based on allied PANI nanofibers has yet to be accomplished, because it is difficult to directly electro-spun PANI nanofibers to fabricate highly conducting materials. Therefore, to conquer this probem, Bon Kang Gu *et al.* developed the polyurethane/PANI (PU-PANI) hybrid-type nanofibrous bundle. This assembly consists of PANI nanofibers (900-nm diameter) that respond to electrical stimulus and responding a precise and high actuation strain (1.65%) at an applied stress (1.03 MPa) in the presence of 1 M CH_4O_3S. The maximum strain generated in the hybrid nanofibrous bundle PANI exhibit elevated electrical conductivity around 0.5 S/cm imparted by PANI. The nanofibrous bundle duplicating artificial muscles behavior and exhibiting more than 75% working efficiency per cycle for the electrochemical bending deformation even after more than 100 cycles. The proposed nanofibrous bundle based on PU/PANI actuated repeatedly without any noticeable creep up to 11 mN (2.263 Mpa) practical load after that considerable creep action can be noticed.

7.4.8   CP as Fibrous Actuator

Electrospinning is readily employed for producing fibers of highly defined shapes, with a wide range of diameters and functionalities, from various types of polymer precursors [133, 134]. However, electrospun fibers of PANI are tricky to fabricate because it is hard to dissolved PANI in biocompatible solvents, and the electrospinning process interferes with its conductivity. Therefore, to get the materials of desired functionalities and properties, a number of fabrication techniques need to be combined. Bhardwaj *et al.* reported a new hybrid-type actuator fabricated by combining sputtering metallization, elecrospinning, and electrochemical deposition of PANI [133, 134]. Thus, aligned Poly(methyl methacrylate) fibers in predetermined conditions were prepared by electrospinning. To get the electrodeposition of PANI polymer in well controlled way, such fibers grow on copper frames, which was coated with gold layer on one side in order to get conductive surface of fibers. Therefore, the metalized side fibers were adopted as working electrode for PANI deposition through electrochemical route. The deposition procedure was carried out in such a way that the fibers were layered with thin film of PANI merely on the side of gold. The actuation behaviors of the fabricated fibers were demonstrated in electrolyte solution 1 M H_2SO_4. The results obtained for this CP composite-based actuator demonstrated that the fibers coated with PANI one side show the maximum displacement (0.1 cm) with actuation time of about 5 s under applied potential of −0.2 and +1.4 V required for contraction and expansion. Whereas, the actuator fibers completely covered with PANI did not show any deformation due to the symmetrical volume changes.

7.4.9   CPs Composite-Based Flapping Wing Actuator

Uh *et al.* reported novel electrolyte-free durable PANI-based actuator, which respond rapidly at low applied voltage [120]. Moreover, in the vicinity of magnetic field, the fabricated PANI-based actuator shows flapping wing motion behavior (Figure 7.3a). This CP-based actuator demonstrated as electrolyte-free, polymer electroactive actuator with single component, showing high durability, fast response time, and operated at low driving

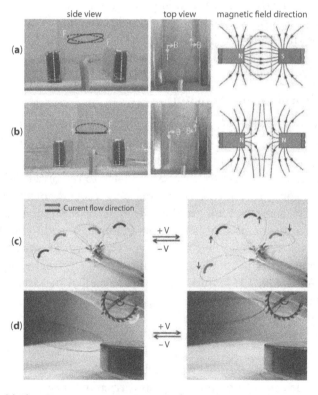

Figure 7.3 (a) Flapping actuator wings positioned among two opposite magnets poles, swings up and down, and the fiber moves in opposite direction near the magnet due to applied current to the polymer fiber. (b) Flapping wings positioned between two same poles of magnets flaps in similar way as shown by arrow. (c) Flapping motion of PANI fiber-based actuator wings. Reversing the directions of current flow results into opposite flapping. (d) Gearwheel rotation by the developed PANI fiber actuator [120].

potential (<5 V). The techniques adopted to fabricate this composite material includes wet-spinning of a pre-organized PANI gel doped with camphor sulfonic acid, which produce flexible, conductive (270 S cm^{-1}), and extended microfibers. The bending motion of this actuator can also be reversed upon alternating the current to the PANI polymer. This actuation, initiated due to a considerably small operating potential (<0.50 V) along with an applied magnetic field, has a extremely great swinging motion (9,000 swings/min) exactly in the similar manner as attain by bees and flies (1,000 to 15,000 swings/min), without any fatigue-resistant up to beyond one million cycles. Firstly, the operating voltage was small and the flapping and bending of the PANI-based fibers occurs under 0.5 V (flapping motion) and 3 V (bending actuation), respectively (Figure 7.3b). Secondly, the flapping speed gained with the PANI-based actuator was analogous to the flapping wings of insect. This trend makes the actuation durable and fatigue-resistant fabricated device attractive as a biomimetic actuator. These observations confirm that the proposed composition has the potential to open new routes for the applications of CP$_S$-based microfibers.

7.4.10   CPs Composite-Based Micro-Gripper System

Inamuddin *et al.* reported a hybrid soft actuator made up of sulfonated poly(vinyl alcohol)/PPy (SPVA/PPy) membrane with platinum metal as electrode, which can be used in microrobotic applications [23]. This membrane actuator was developed by combining solution casting method along with electroless plating technique, where SPVA/PPy membrane was used as electrolyte material. Electroding of Pt metal was carried out over SPVA/PPy membrane surfaces to fabricate SPVA/PPy/Pt polymer membrane actuator. After developing the actuator membranes, they developed a compliant three finger-based gripping device with a computer control system using SPVAPPy/Pt membranes (Figure 7.4). In the developed microgripping device, all three fingers were made up of SPVA/PPy/Pt actuator membranes. Here, all three fabricated polymer actuator-based fingers were fixed in a wrist, which was integrated with holder. To operate the microgripper, electric potential (0–5.25 VDC) was applied *via* proportional-derivative controller. Under applied potential, all the three SPVA/PPy/Pt-based fingers bend together and grasp the object. After that, when voltage was released, the SPVA/PPy/Pt-based actuator fingers bends in opposite manner and release the object.

The maximum bending deformation of the fabricated PPy-based membrane actuator was found to be 18.5-mm along with the load carrying ability up to 1.9 mN. The extensive tip deformation with flexible nature

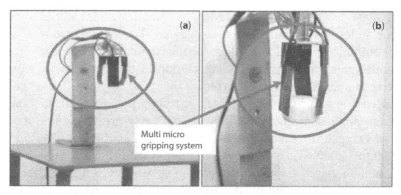

(a) Microgripper without object   (b) Microgripper with object

Figure 7.4 Three-finger–based microgripping device developed through PPy composite-based ionic polymer membrane actuator [23].

and considerable tip generating force abilities are appropriate for handling miniature objects. The operation of actuator fingers was carried out by electrical stimulated CP composite membranes as an alternative of conventional motor. Each fabricated actuator finger was capable to undergo bending deformation individually for dexterous handling. Therefore, this kind of CP composite-based actuator sounds to be enormously constructive for lifting light weight tiny objects and can be used in micro robotics and other biomemetic applications.

7.5   Summary and Future Perspective of CPs Composites-Based Actuators

As the finest studied and potentially most practical and constructive CPs, e.g., PANI and PPy, both had a broad area of applications due to its unique doping/dedoping, inherent conductivity, and facile synthesis. CPs (such as PPy and PANI) and their composite materials produced in the pursuit of CPs applications such as sensors, actuators, bioinspired robotics and artificial muscles, biomedical products, biotechnology, optoelectronics drug delivery, adhesives, waterborne paints, sensitive adhesives, coatings, and magnetic, etc., are supposed to appear as versatile technology in various fields. The development of novel soft actuators and sensors will be investigated with different strategies that utilize CPs as sensing and actuating elements to discover small organic, inorganic, bio-molecules, and used as artificial muscles for drug delivery. Heteroatom doped PANI or PPy can be utilized to fabricate novel microgrippers, flapping wings, and lot of research will be carried out on CPs composites and their nanomaterials. Biocompatibility of CPs and their nanocomposites in different robotics and biomemetics applications will need to be addressed for further future works. No matter how many types of applications are discussed here, there are forever many more thoughts that can be accomplished with the advancement of novel materials particularly that can be constructed cost-effectively at massive scale. The mass construction of single-layer PPy or PANI is a main confront that needs to be sort it out in the future. Overall, the CPs and their composites will come out as important players in new generation soft actuator, electronics devices, and modern medicine delivery system, etc. Over electrochemical and interfacial polymerization or by rapid mixing technique, the nanofiber morphology of PANI can be easily achieved. Fabricated devices by means of nanostructured PANI generally delivered improved performance related to the devices made up of using their usual counterparts. For example, PANI nanofibers-based actuators exhibit quick response and superior repeatability; while surface electrodes made with PANI nanofibers show better cycle stability and higher specific capacitance. PANI nanofibers easily disperse in solvents at suitable pH, which facilitate them to easily incorporate into matrices for electromagnetic interference and electrostatic dissipative shielding. Furthermore, the joint actuation of CPs composites may assist the fabrication of actuators with new properties. As result of this perception, a walking device containing PPy composite with two legs and one arm was manufactured [66]. This device shows the delivery capability and controlled transportation under NIR light and humidity actuation. This type of observation not only shows a path to think about manifold actuation of the CPs composite-based actuators, although also suggests the route to fabricate promising multi- responsive soft actuator for enhanced-performance bioinspired robotics and artificial muscles.

Acknowledgements

The authors are grateful to the Department of Chemistry, King Abdulaziz University, Saudi Arabia for providing research facilities and granting the permission to publish this work. This work was supported by Ministry of Education, King Abdulaziz University Administration of Support for Research and Development Initiatives, Kingdom of Saudi Arabia, under the research scheme Post-Doctoral Researcher awarded to Dr. Ajahar Khan.

References

1. Ma, M., Guo, L., Anderson, D.G., Langer, R., Bio-Inspired Polymer Composite Actuator and Generator Driven by Water Gradients. *Science* (80-.), 339, 186–189, 2013, doi: 10.1126/science.1230262.

2. Mu, J., Hou, C., Zhu, B., Wang, H., Li, Y., Zhang, Q., A multi-responsive water-driven actuator with instant and powerful performance for versatile applications. *Sci. Rep.*, 5, 9503, 2015, doi: 10.1038/srep09503.

3. Wei, J. and Yu, Y., Photodeformable polymer gels and crosslinked liquid-crystalline polymers. *Soft Matter.*, 8, 8050, 2012, doi: 10.1039/c2sm25474c.

4. Jiang, H.Y., Kelch, S., Lendlein, A., Polymers Move in Response to Light. *Adv. Mater.*, 18, 1471–1475, 2006, doi: 10.1002/adma.200502266.

5. Li, J., Ma, W., Song, L., Niu, Z., Cai, L., Zeng, Q., Zhang, X., Dong, H., Zhao, D., Zhou, W., Xie, S., Superfast-Response and Ultrahigh-Power-Density Electromechanical Actuators Based on Hierarchal Carbon Nanotube Electrodes and Chitosan. *Nano Lett.*, 11, 4636–4641, 2011, doi: 10.1021/nl202132m.

6. Kong, L. and Chen, W., Carbon Nanotube and Graphene-based Bioinspired Electrochemical Actuators. *Adv. Mater.*, 26, 1025–1043, 2014, doi: 10.1002/adma.201303432.

7. Stroganov, V., Zakharchenko, S., Sperling, E., Meyer, A.K., Schmidt, O.G., Ionov, L., Biodegradable Self-Folding Polymer Films with Controlled Thermo-Triggered Folding. *Adv. Funct. Mater.*, 24, 4357–4363, 2014, doi: 10.1002/adfm.201400176.

8. Ward, M.A. and Georgiou, T.K., Thermoresponsive Polymers for Biomedical Applications. *Polymers (Basel)*, 3, 1215–1242, 2011, doi: 10.3390/polym3031215.

9. Henn, D.M., Fu, W., Mei, S., Li, C.Y., Zhao, B., Temperature-Induced Shape Changing of Thermosensitive Binary Heterografted Linear Molecular Brushes between Extended Wormlike and Stable Globular Conformations. *Macromolecules*, 50, 1645–1656, 2017, doi: 10.1021/acs.macromol.7b00150.

10. Chen, J.-K. and Chang, C.-J., Fabrications and Applications of Stimulus-Responsive Polymer Films and Patterns on Surfaces: A Review. *Materials (Basel)*, 7, 805–875, 2014, doi: 10.3390/ma7020805.

11. Jung, J.-Y. and Oh, I.-K., Novel Nanocomposite Actuator Based on Sulfonated Poly(styrene-b-ethylene-co-butylene-b-styrene) Polymer. *J. Nanosci. Nanotechnol.*, 7, 3740–3743, 2007, doi: 10.1166/jnn.2007.004.

12. Lu, J., Kim, S.-G., Lee, S., Oh, I.-K., Fabrication and actuation of electro-active polymer actuator based on PSMI-incorporated PVDF. *Smart Mater. Struct.*, 17, 045002, 2008, doi: 10.1088/0964-1726/17/4/045002.

13. Ikeda, K., Sasaki, M., Tamagawa, H., IPMC bending predicted by the circuit and viscoelastic models considering individual influence of Faradaic and non-Faradaic currents on the bending. *Sens. Actuators B Chem.*, 190, 954–967, 2014, doi: 10.1016/j.snb.2013.09.016.

14. M'boungui, G., Semail, B., Giraud, F., Jimoh, A.A.A., Development of a novel plane piezoelectric actuator using Hamilton's principle based model and Hertz contact theory. *Sens. Actuators A Phys.*, 217, 116–123, 2014, doi: 10.1016/j.sna.2014.06.026.

15. Qin, C., Feng, Y., An, H., Han, J., Cao, C., Feng, W., Tetracarboxylated Azobenzene/Polymer Supramolecular Assemblies as High-Performance Multiresponsive Actuators. *ACS Appl. Mater. Interfaces*, 9, 4066–4073, 2017, doi: 10.1021/acsami.6b15075.

16. Zhang, Q., Yu, Y., Yang, K., Zhang, B., Zhao, K., Xiong, G., Zhang, X., Mechanically robust and electrically conductive graphene-paper/glass-fibers/epoxy composites for stimuli-responsive sensors and Joule heating deicers. *Carbon N. Y.*, 124, 296–307, 2017, doi: 10.1016/j.carbon.2017.09.001.

17. Liakos, I.L., Mondini, A., Filippeschi, C., Mattoli, V., Tramacere, F., Mazzolai, B., Towards ultra-responsive biodegradable polysaccharide humidity sensors. *Mater. Today Chem.*, 6, 1–12, 2017, doi: 10.1016/j.mtchem.2017.08.001.

18. Chen, M., Frueh, J., Wang, D., Lin, X., Xie, H., He, Q., Polybenzoxazole Nanofiber-Reinforced Moisture-Responsive Soft Actuators. *Sci. Rep.*, 7, 769, 2017, doi: 10.1038/s41598-017-00870-w.

19. Wang, Z., Wang, W., Yu, D., Pressure responsive PET fabrics via constructing conductive wrinkles at room temperature. *Chem. Eng. J.*, 330, 146–156, 2017, doi: 10.1016/j.cej.2017.07.094.

20. Si, Y., Wang, L., Wang, X., Tang, N., Yu, J., Ding, B., Ultrahigh-Water-Content, Superelastic, and Shape-Memory Nanofiber-Assembled Hydrogels Exhibiting Pressure-Responsive Conductivity. *Adv. Mater.*, 29, 1700339, 2017, doi: 10.1002/adma.201700339.

21. Wang, P.-C., Chao, C.-I., Lin, W.-K., Hung, S.-Y., All-polymer variable resistors as pressure-responsive devices photochemically assembled using porous composites and flexible electrodes based on conducting polymers. *J. Chin. Inst. Eng.*, 35, 595–599, 2012, doi: 10.1080/02533839.2012.679113.

22. Inamuddin, Khan, A., Jain, R.K., Naushad, M., Study and preparation of highly water-stable polyacrylonitrile-kraton-graphene composite membrane for bending actuator toward robotic application. *J. Intell. Mater. Syst. Struct.*, 27, 1534–1546, 2016, doi: 10.1177/1045389X15596627.

23. Inamuddin, Khan, A., Jain, R.K., Naushad, M., Development of sulfonated poly(vinyl alcohol)/polpyrrole based ionic polymer metal composite (IPMC) actuator and its characterization. *Smart Mater. Struct.*, 24, 95003, 2015, doi. org/10.1088/0964-1726/24/9/095003

24. Cui, Z., Zhou, M., Greensmith, P.J., Wang, W., Hoyland, J.A., Kinloch, I.A., Freemont, T., Saunders, B.R., A study of conductive hydrogel composites of pH-responsive microgels and carbon nanotubes. *Soft Matter*, 12, 4142–4153, 2016, doi: 10.1039/C6SM00223D.

25. Ahmad, H., Rahman, M.M., Ali, M.A., Minami, H., Tauer, K., Gafur, M.A., Rahman, M.M., A simple route to synthesize conductive stimuli-responsive polypyrrole nanocomposite hydrogel particles with strong magnetic

properties and their performance for removal of hexavalent chromium ions from aqueous solution. *J. Magn. Magn. Mater.*, 412, 15–22, 2016, doi: 10.1016/j.jmmm.2016.03.068.

26. Abu-Thabit, N., Umar, Y., Ratemi, E., Ahmad, A., Ahmad Abuilaiwi, F., A Flexible Optical pH Sensor Based on Polysulfone Membranes Coated with pH-Responsive Polyaniline Nanofibers. *Sensors*, 16, 986, 2016, doi: 10.3390/s16070986.

27. Grigoryev, A., Sa, V., Gopishetty, V., Tokarev, I., Kornev, K.G., Minko, S., Wet-Spun Stimuli-Responsive Composite Fibers with Tunable Electrical Conductivity. *Adv. Funct. Mater.*, 23, 5903–5909, 2013, doi: 10.1002/adfm.201203721.

28. Boruah, M., Kalita, A., Pokhrel, B., Dolui, S.K., Boruah, R., Synthesis and Characterization of pH Responsive Conductive Composites of Poly(acrylic acid-co-acrylamide) Impregnated with Polyaniline by Interfacial Polymerization. *Adv. Polym. Technol.*, 32, E520–E530, 2013, doi: 10.1002/adv.21298.

29. Rus, D. and Tolley, M.T., Design, fabrication and control of soft robots. *Nature*, 521, 467, 2015, doi: 10.1038/nature14543.

30. Cianchetti, M., Laschi, C., Menciassi, A., Dario, P., Biomedical applications of soft robotics. *Nat. Rev. Mater.*, 3, 143–153, 2018, doi: 10.1038/s41578-018-0022-y.

31. Zhang, Q.M., Giant Electrostriction and Relaxor Ferroelectric Behavior in Electron-Irradiated Poly(vinylidene fluoride-trifluoroethylene) Copolymer. *Science (80-.)*, 280, 2101–2104, 1998, doi: 10.1126/science.280.5372.2101.

32. Rosset, S. and Shea, H.R., Flexible and stretchable electrodes for dielectric elastomer actuators. *Appl. Phys. A*, 110, 281–307, 2013, doi: 10.1007/s00339-012-7402-8.

33. Jun, K., Kim, D., Ryu, S., Oh, I.-K., Surface Modification of Anisotropic Dielectric Elastomer Actuators with Uni- and Bi-axially Wrinkled Carbon Electrodes for Wettability Control. *Sci. Rep.*, 7, 6091, 2017, doi: 10.1038/s41598-017-06274-0.

34. Shahinpoor, M. and Kim, K.J., Ionic polymer–metal composites: IV. Industrial and medical applications. *Smart Mater. Struct.*, 14, 197, 2005, doi. org/10.1088/0964-1726/14/1/020

35. Park, H., Temenoff, J.S., Tabata, Y., Caplan, A.I., Mikos, A.G., Injectable biodegradable hydrogel composites for rabbit marrow mesenchymal stem cell and growth factor delivery for cartilage tissue engineering. *Biomaterials*, 28, 3217–3227, 2007, doi: 10.1016/j.biomaterials.2007.03.030.

36. Nakabo Y., Mukai T., Asaka K., Biomimetic Soft Robots Using IPMC. In: *Electroactive Polymers for Robotic Applications*, K.J. Kim and S. Tadokoro (eds). Springer, London, 2007, doi.org/10.1007/978-1-84628-372-7_7.

37. Yeom, S.-W. and Oh, I.-K., A biomimetic jellyfish robot based on ionic polymer metal composite actuators. *Smart Mater. Struct.*, 18, 085002, 2009, doi: 10.1088/0964-1726/18/8/085002.

38. Aureli, M., Kopman, V., Porfiri, M., Free-Locomotion of Underwater Vehicles Actuated by Ionic Polymer Metal Composites. *IEEE/ASME Trans. Mechatron.*, 15, 603–614, 2010, doi: 10.1109/TMECH.2009.2030887.

39. Khan, A., Jain, R.K., Banerjee, P., Inamuddin, Asiri, A.M., Soft actuator based on Kraton with GO/Ag/Pani composite electrodes for robotic applications. *Mater. Res. Express*, 4, 115701, 2017, doi: 10.1088/2053-1591/aa9394.

40. Otero, T.F., Angulo, E., Rodríguez, J., Santamaría, C., Electrochemomechanical properties from a bilayer: Polypyrrole/non-conducting and flexible material—Artificial muscle. *J. Electroanal. Chem.*, 341, 369–375, 1992, doi: 10.1016/022-0728(92)80495-P.

41. Otero, B.T.F. and Sansin, J.M., Soft and Wet Conducting Polymers for Artificial Muscles Soft and Wet Conducting Polymers for Artificial Muscles **. *Adv. Mater.*, 4095, 491–494, 2015, doi: 10.1002/(SICI)1521-4095(199804)10.

42. Guimard, N.K., Gomez, N., Schmidt, C.E., Conducting polymers in biomedical engineering. *Prog. Polym. Sci.*, 32, 876–921, 2007, doi: 10.1016/j.progpolymsci.2007.05.012.

43. Smela, E., Conjugated Polymer Actuators for Biomedical Applications. *Adv. Mater.*, 15, 481–494, 2003, doi: 10.1002/adma.200390113.

44. Baughman, R.H., Conducting polymer artificial muscles. *Synth. Met.*, 78, 339–353, 1996, doi: 10.1016/0379-6779(96)80158-5.

45. Beregoi, M., Busuioc, C., Evanghelidis, A., Matei, E., Iordache, F., Radu, M., Dinischiotu, A., Enculescu, I., Electrochromic properties of polyaniline-coated fiber webs for tissue engineering applications. *Int. J. Pharm.*, 510, 465–473, 2016, doi: 10.1016/j.ijpharm.2015.11.055.

46. Ansari, R. and Keivani, M.B., Polyaniline Conducting Electroactive Polymers Thermal and Environmental Stability Studies. *E-J. Chem.*, 3, 202–217, 2006, doi: 10.1155/2006/395391.

47. Pei, Q., Inganas, O., Lundstrom, I., Bending bilayer strips built from polyaniline for artificial electrochemical muscles. *Smart Mater. Struct.*, 2, 1–6, 1993, doi: 10.1088/0964-1726/2/1/001.

48. Jager, E.W.H., Microfabricating Conjugated Polymer Actuators. *Science (80-.)*, 290, 1540–1545, 2000, doi: 10.1126/science.290.5496.1540.

49. Madden, J.D., Cush, R.A., Kanigan, T.S., Hunter, I.W., Fast contracting polypyrrole actuators. *Synth. Met.*, 113, 185–192, 2000, doi: 10.1016/S0379-6779(00)00195-8.

50. Madden, J.D.W., Vandesteeg, N.A., Anquetil, P.A., Madden, P.G.A., Takshi, A., Pytel, R.Z., Lafontaine, S.R., Wieringa, P.A., Hunter, I.W., Artificial Muscle Technology: Physical Principles and Naval Prospects. *IEEE J. Ocean. Eng.*, 29, 706–728, 2004, doi: 10.1109/JOE.2004.833135.

51. Xi, B., Truong, V.-T., Mottaghitalab, V., Whitten, P.G., Spinks, G.M., Wallace, G.G., Actuation behaviour of polyaniline films and tubes prepared by the phase inversion technique. *Smart Mater. Struct.*, 16, 1549–1554, 2007, doi: 10.1088/0964-1726/16/5/007.

52. Qi, B., Lu, W., Mattes, B.R., Strain and Energy Efficiency of Polyaniline Fiber Electrochemical Actuators in Aqueous Electrolytes. *J. Phys. Chem. B*, 108, 6222–6227, 2004, doi: 10.1021/jp031092s.

53. Sansiñena, J.-M., Gao, J., Wang, H.-L., High-Performance, Monolithic Polyaniline Electrochemical Actuators. *Adv. Funct. Mater.*, 13, 703–709, 2003, doi: 10.1002/adfm.200304347.

54. Gu, B.K., Ismail, Y.A., Spinks, G.M., Kim, S.I., So, I., Kim, S.J., A Linear Actuation of Polymeric Nanofibrous Bundle for Artificial Muscles. *Chem. Mater.*, 21, 511–515, 2009, doi: 10.1021/cm802377d.

55. Smela, E., Inganäs, O., Pei, Q., Lundström, I., Electrochemical muscles: Micromachining fingers and corkscrews. *Adv. Mater.*, 5, 630–632, 1993, doi: 10.1002/adma.19930050905.

56. Smela, E., Inganas, O., Lundstrom, I., Controlled Folding of Micrometer-Size Structures. *Science (80-.)*, 268, 1735–1738, 1995, doi: 10.1126/science.268.5218.1735.

57. José, L., *Pons, Emerging Actuator Technologies: A Micromechatronic Approach*, John Wiley & Sons, Suxxes, England, 2005.

58. Bar-Cohen, Y., *Electroactive polymer (EAP) actuators as artificial muscles: Reality, potential, and challenges*, 2nd ed, SPIE Publications, Bellingham, Washington, 2004, 9780819452979.

59. Mirfakhrai, T., Madden, J.D.W., Baughman, R.H., Polymer artificial muscles. *Mater. Today*, 10, 30–38, 2007, doi: 10.1016/S1369-7021(07)70048-2.

60. Hara, S., Zama, T., Takashima, W., Kaneto, K., Tris(trifluoromethylsulfonyl) methide-doped polypyrrole as a conducting polymer actuator with large electrochemical strain. *Synth. Met.*, 156, 351–355, 2006, doi: 10.1016/j.synthmet.2006.01.001.

61. Bay, L., West, K., Sommer-Larsen, P., Skaarup, S., Benslimane, M., A Conducting Polymer Artificial Muscle with 12% Linear Strain. *Adv. Mater.*, 15, 310–313, 2003, doi: 10.1002/adma.200390075.

62. Schuhmann, W., Lammert, R., Hämmerle, M., Schmidt, H.-L., Electrocatalytic properties of polypyrrole in amperometric electrodes. *Biosens. Bioelectron.*, 6, 689–697, 1991, doi: 10.1016/0956-5663(91)87023-5.

63. Coche-Guerente, L., Cosnier, S., Innocent, C., Mailley, P., Moutet, J.-C., Morélis, R.M., Leca, B., Coulet, P.R., Controlled electrochemical preparation of enzymatic layers for the design of amperometric biosensors. *Electroanalysis*, 5, 647–652, 1993, doi: 10.1002/elan.1140050804.

64. Khan, A., Inamuddin, Jain, R.K., Easy, operable ionic polymer metal composite actuator based on a platinum-coated sulfonated poly(vinyl alcohol)-polyaniline composite membrane. *J. Appl. Polym. Sci.*, 133, 43787, 2016, doi. org/10.1002/app.43787.

65. Sabouraud, G., Sadki, S., Brodie, N., The mechanisms of pyrrole electropolymerization. *Chem. Soc. Rev.*, 29, 283–293, 2000, doi: 10.1039/a807124a.

66. Wang, T., Li, M., Zhang, H., Sun, Y., Dong, B., A multi-responsive bidirectional bending actuator based on polypyrrole and agar nanocomposites. *J. Mater. Chem. C*, 6, 6416–6422, 2018, doi: 10.1039/C8TC00747K.

67. Angeli, A. and Alessandri, L., The electrochemistry of conducting polymers: Gazz. Chim. Ital, 46, 279–283, 1916.

68. Oh, E.J., Jang, K.S., MacDiarmid, A.G., High molecular weight soluble polypyrrole. *Synth. Met.*, 125, 267–272, 2001, doi: 10.1016/S0379-6779(01)00384-8.

69. Henry, M.C., Hsueh, C.-C., Timko, B.P., Freund, M.S., Reaction of Pyrrole and Chlorauric Acid A New Route to Composite Colloids. *J. Electrochem. Soc.*, 148, D155, 2001, doi: 10.1149/1.1405802.

70. Schuhmann, W., Lammert, R., Uhe, B., Schmidt, H.-L., Polypyrrole, a new possibility for covalent binding of oxidoreductases to electrode surfaces as a base for stable biosensors. *Sens. Actuators B Chem.*, 1, 537–541, 1990, doi: 10.1016/0925-4005(90)80268-5.

71. Fang, Q., Chetwynd, D.G., Gardner, J.W., Conducting polymer films by UV-photo processing. *Sens. Actuators A Phys.*, 99, 74–77, 2002, doi: 10.1016/S0924-4247(01)00894-9.

72. Deore, B., Chen, Z., Nagaoka, T., Overoxidized Polypyrrole with Dopant Complementary Cavities as a New Molecularly Imprinted Polymer Matrix. *Anal. Sci.*, 15, 827–828, 1999, doi: 10.2116/analsci.15.827.

73. Ramanavičius, A., Kaušaitė, A., Ramanavičienė, A., Polypyrrole-coated glucose oxidase nanoparticles for biosensor design. *Sens. Actuators B Chem.*, 111–112, 532–539, 2005, doi: 10.1016/j.snb.2005.03.038.

74. Liang, H.-J., Ling, T.-R., Rick, J.F., Chou, T.-C., Molecularly imprinted electrochemical sensor able to enantroselectivly recognize d and l-tyrosine. *Anal. Chim. Acta*, 542, 83–89, 2005, doi: 10.1016/j.aca.2005.02.007.

75. Trojanowicz, M. and Wcisło, M., Electrochemical and Piezoelectric Enantioselective Sensors and Biosensors. *Anal. Lett.*, 38, 523–547, 2005, doi: 10.1081/AL-200050157.

76. Ebarvia, B., Cabanilla, S., Sevilla, F., III, Biomimetic properties and surface studies of a piezoelectric caffeine sensor based on electrosynthesized polypyrrole. *Talanta*, 66, 145–152, 2005, doi: 10.1016/j.talanta.2004.10.009.

77. Ramanaviciene, A. and Ramanavicius, A., Molecularly imprinted polypyrrole-based synthetic receptor for direct detection of bovine leukemia virus glycoproteins. *Biosens. Bioelectron.*, 20, 1076–1082, 2004, doi: 10.1016/j.bios.2004.05.014.

78. de Barros, R.A., de Azevedo, W.M., de Aguiar, F.M., Photo-induced polymerization of polyaniline. *Mater. Charact.*, 50, 131–134, 2003, doi: 10.1016/S1044-5803(03)00080-9.

79. Pokrop, R., Zagórska, M., Kulik, M., Kulszewicz-Bajer, I., Dufour, B., Rannou, P., Pron, A., Gondek, E., Sanetra, J., Solution Processible Sulfosuccinate Doped Polypyrrole: Preparation, Spectroscopic and Spectroelectrochemical Characterization. *Mol. Cryst. Liq. Cryst.*, 415, 93–104, 2004, doi: 10.1080/15421400490482934.

80. Schmidt, H.-L., Gutberlet, F., Schuhmann, W., New principles of amperometric enzyme electrodes and of reagentless oxidoreductase biosensors. *Sens. Actuators B Chem.*, 13, 366–371, 1993, doi: 10.1016/0925-4005(93)85403-W.

81. Kuramoto, N. and Tomita, A., Chemical oxidative polymerization of dodecylbenzenesulfonic acid aniline salt in chloroform. *Synth. Met.*, 88, 147–151, 1997, doi: 10.1016/S0379-6779(97)03850-2.
82. Miras, M.C., Barbero, C., Haas, O., Preparation of polyaniline by electrochemical polymerization of aniline in acetonitrile solution. *Synth. Met.*, 43, 3081–3084, 1991, doi: 10.1016/0379-6779(91)91243-4.
83. Dallas, P., Stamopoulos, D., Boukos, N., Tzitzios, V., Niarchos, D., Petridis, D., Characterization, magnetic and transport properties of polyaniline synthesized through interfacial polymerization. *Polymer (Guildf)*, 48, 3162–3169, 2007, doi: 10.1016/j.polymer.2007.03.055.
84. Chen, J., Chao, D., Lu, X., Zhang, W., Novel interfacial polymerization for radially oriented polyaniline nanofibers. *Mater. Lett.*, 61, 1419–1423, 2007, doi: 10.1016/j.matlet.2006.07.043.
85. Guo, Q., Yi, C., Zhu, L., Yang, Q., Xie, Y., Chemical synthesis of cross-linked polyaniline by a novel solvothermal metathesis reaction of p-dichlorobenzene with sodium amide. *Polymer (Guildf)*, 46, 3185–3189, 2005, doi: 10.1016/j.polymer.2005.01.092.
86. Yang, C.-H., Huang, L.-R., Chih, Y.-K., Lin, W.-C., Liu, F.-J., Wang, T.-L., Molecular assembled self-doped polyaniline copolymer ultrathin films. *Polymer (Guildf)*, 48, 3237–3247, 2007, doi: 10.1016/j.polymer.2007.04.013.
87. Kim, J., Kim, E., Won, Y., Lee, H., Suh, K., The preparation and characteristics of conductive poly(3,4-ethylenedioxythiophene) thin film by vapor-phase polymerization. *Synth. Met.*, 139, 485–489, 2003, doi: 10.1016/S0379-6779(03)00202-9.
88. Kim, J., Kwon, S., Han, S., Min, Y., Nanofilms Based on Vapor Deposition of Polymerized Polypyrrole and its Characteristics. *Jpn. J. Appl. Phys.*, 43, 5660–5664, 2004, doi: 10.1143/JJAP.43.5660.
89. Kim, J.-Y., Lee, J.-H., Kwon, S.-J., The manufacture and properties of polyaniline nano-films prepared through vapor-phase polymerization. *Synth. Met.*, 157, 336–342, 2007, doi: 10.1016/j.synthmet.2007.03.013.
90. Zhang, L. and Wan, M., Chiral polyaniline nanotubes synthesized via a self-assembly process. *Thin Solid Films*, 477, 24–31, 2005, doi: 10.1016/j.tsf.2004.08.106.
91. Jing, X., Wang, Y., Wu, D., She, L., Guo, Y., Polyaniline nanofibers prepared with ultrasonic irradiation. *J. Polym. Sci. Part A Polym. Chem.*, 44, 1014–1019, 2006, doi: 10.1002/pola.21217.
92. Jing, X., Wang, Y., Wu, D., Qiang, J., Sonochemical synthesis of polyaniline nanofibers. *Ultrason. Sonochem.*, 14, 75–80, 2007, doi: 10.1016/j.ultsonch.2006.02.001.
93. J., L. and Diaz, A.F., Electroactive polyaniline films. *J. Electroanal. Chem.*, 111, 111–114, 1980, doi.org/10.1016/S0022-0728(80)80081-7

94. Genies, E.M., Syed, A.A., Tsintavis, C., Electrochemical Study Of Polyaniline In Aqueous And Organic Medium. Redox And Kinetic Properties. *Mol. Cryst. Liq. Cryst.*, 121, 181–186, 1985, doi: 10.1080/00268948508074858.

95. Genies, E.M. and Tsintavis, C., Redox mechanism and electrochemical behaviour or polyaniline deposits. *J. Electroanal. Chem. Interfacial Electrochem.*, 195, 109–128, 1985, doi: 10.1016/0022-0728(85)80009-7.

96. Bhadra, S., Chattopadhyay, S., Singha, N.K., Khastgir, D., Improvement of conductivity of electrochemically synthesized polyaniline. *J. Appl. Polym. Sci.*, 108, 57–64, 2008, doi: 10.1002/app.26926.

97. Mengoli, G., Munari, M.-T., Folonari, C., Anodic formation of polynitroanilide films onto copper. *J. Electroanal. Chem. Interfacial Electrochem.*, 124, 237–246, 1981, doi: 10.1016/S0022-0728(81)80301-4.

98. Mengoli, G., Munari, M.T., Bianco, P., Musiani, M.M., Anodic synthesis of polyaniline coatings onto fe sheets. *J. Appl. Polym. Sci.*, 26, 4247–4257, 1981, doi: 10.1002/app.1981.070261224.

99. Camalet, J.L., Lacroix, J.C., Aeiyach, S., Chane-Ching, K., Lacaze, P.C., Electrosynthesis of adherent polyaniline films on iron and mild steel in aqueous oxalic acid medium. *Synth. Met.*, 93, 133–142, 1998, doi: 10.1016/S0379-6779(97)04099-X.

100. DeBerry, D.W., Modification of the Electrochemical and Corrosion Behavior of Stainless Steels with an Electroactive Coating. *J. Electrochem. Soc.*, 132, 1022, 1985, doi: 10.1149/1.2114008.

101. Bhadra, S., Singha, N.K., Chattopadhyay, S., Khastgir, D., Effect of different reaction parameters on the conductivity and dielectric properties of polyaniline synthesized electrochemically and modeling of conductivity against reaction parameters through regression analysis. *J. Polym. Sci. Part B Polym. Phys.*, 45, 2046–2059, 2007, doi: 10.1002/polb.21175.

102. Bhadra, S., Singha, N.K., Khastgir, D., Dual functionality of PTSA as electrolyte and dopant in the electrochemical synthesis of polyaniline, and its effect on electrical properties. *Polym. Int.*, 56, 919–927, 2007, doi: 10.1002/pi.2225.

103. Bhadra, S., Singha, N.K., Khastgir, D., Electrochemical synthesis of polyaniline and its comparison with chemically synthesized polyaniline. *J. Appl. Polym. Sci.*, 104, 1900–1904, 2007, doi: 10.1002/app.25867.

104. Paul, E.W., Ricco, A.J., Wrighton, M.S., Resistance of polyaniline films as a function of electrochemical potential and the fabrication of polyaniline-based microelectronic devices. *J. Phys. Chem.*, 89, 1441–1447, 1985, doi: 10.1021/j100254a028.

105. Hussain, A.M.P. and Kumar, A., Electrochemical synthesis and characterization of chloride doped polyaniline. *Bull. Mater. Sci.*, 26, 329–334, 2003, doi: 10.1007/BF02707455.

106. Iroh, J.O., Zhu, Y., Shah, K., Levine, K., Rajagopalan, R., Uyar, T., Donley, M., Mantz, R., Johnson, J., Voevodin, N.N., Balbyshev, V., Khramov,

A., Electrochemical synthesis: A novel technique for processing multi-functional coatings. *Prog. Org. Coat.*, 47, 365–375, 2003, doi: 10.1016/j. porgcoat.2003.07.006.

107. Hand, R.L. and Nelson, R.F., Anodic oxidation pathways of N-alkylanilines. *J. Am. Chem. Soc.*, 96, 850–860, 1974, doi: 10.1021/ja00810a034.

108. Hand, R.L., The Anodic Decomposition Pathways of Ortho- and Meta-substituted Anilines. *J. Electrochem. Soc.*, 125, 1059, 1978, doi: 10.1149/1.2131621.

109. Gupta, V. and Miura, N., High performance electrochemical supercapacitor from electrochemically synthesized nanostructured polyaniline. *Mater. Lett.*, 60, 1466–1469, 2006, doi: 10.1016/j.matlet.2005.11.047.

110. Kobayashi, N., Teshima, K., Hirohashi, R., Conducting polymer image formation with photoinduced electron transfer reaction. *J. Mater. Chem.*, 8, 497–506, 1998, doi: 10.1039/a706386e.

111. Khanna, P.K., Singh, N., Charan, S., Viswanath, A.K., Synthesis of Ag/poly-aniline nanocomposite via an *in situ* photo-redox mechanism. *Mater. Chem. Phys.*, 92, 214–219, 2005, doi: 10.1016/j.matchemphys.2005.01.011.

112. Baker, C.O., Huang, X., Nelson, W., Kaner, R.B., Polyaniline nanofibers: Broadening applications for conducting polymers. *Chem. Soc. Rev.*, 46, 1510–1525, 2017, doi: 10.1039/C6CS00555A.

113. Baker, C.O., Shedd, B., Innis, P.C., Whitten, P.G., Spinks, G.M., Wallace, G.G., Kaner, R.B., Monolithic Actuators from Flash-Welded Polyaniline Nanofibers. *Adv. Mater.*, 20, 155–158, 2008, doi: 10.1002/adma.200602864.

114. Gao, H., Zhang, J., Yu, W., Li, Y., Zhu, S., Li, Y., Wang, T., Yang, B., Monolithic polyaniline/polyvinyl alcohol nanocomposite actuators with tunable stimuli-responsive properties. *Sens. Actuators B Chem.*, 145, 839–846, 2010, doi: 10.1016/j.snb.2010.01.066.

115. Shedd, B., Baker, C.O., Heller, M.J., Kaner, R.B., Hahn, H.T., Fabrication of monolithic microstructures from polyaniline nanofibers. *Mater. Sci. Eng. B*, 162, 111–115, 2009, doi: 10.1016/j.mseb.2009.03.012.

116. Xu, H., Bharti, V., Cheng, Z.-Y., Zhang, Q.M., Conduction Behavior of Doped Polyaniline Under High Current Density and the Performance of an all Polymer Electromechanical System. *MRS Proc.*, 600, 185, 1999, doi: 10.1557/PROC-600-185.

117. Molberg, M., Crespy, D., Rupper, P., Nüesch, F., Månson, J.-A.E., Löwe, C., Opris, D.M., High Breakdown Field Dielectric Elastomer Actuators Using Encapsulated Polyaniline as High Dielectric Constant Filler. *Adv. Funct. Mater.*, 20, 3280–3291, 2010, doi: 10.1002/adfm.201000486.

118. Wang, D., Zhang, L., Zhang, L., Zha, J.-W., Yu, H., Hu, S., Dang, Z.-M., Enhanced electro-mechanical actuation strain in polyaniline nanorods/silicone rubber nanodielectric elastomer films. *Appl. Phys. Lett.*, 104, 242903, 2014, doi: 10.1063/1.4884362.

119. Kim, S.H., Oh, K.W., Choi, J.H., Preparation and self-assembly of polyaniline nanorods and their application as electroactive actuators. *J. Appl. Polym. Sci.*, 116, 2601–2609, 2010, doi: 10.1002/app.31782.

120. Uh, K., Yoon, B., Lee, C.W., Kim, J.-M., An Electrolyte-Free Conducting Polymer Actuator that Displays Electrothermal Bending and Flapping Wing Motions under a Magnetic Field. *ACS Appl. Mater. Interfaces*, 8, 1289–1296, 2016, doi: 10.1021/acsami.5b09981.

121. Beregoi, M., Preda, N., Evanghelidis, A., Costas, A., Enculescu, I., Versatile Actuators Based on Polypyrrole-Coated Metalized Eggshell Membranes. *ACS Sustain. Chem. Eng.*, 6, 10173–10181, 2018, doi: 10.1021/acssuschemeng.8b01489.

122. Liu, Q., Liu, L., Xie, K., Meng, Y., Wu, H., Wang, G., Dai, Z., Wei, Z., Zhang, Z., Synergistic effect of a r-GO/PANI nanocomposite electrode based air working ionic actuator with a large actuation stroke and long-term durability. *J. Mater. Chem. A*, 3, 8380–8388, 2015, doi: 10.1039/C5TA00669D.

123. Gao, J., Sansiñena, J.-M., Wang, H.-L., Tunable Polyaniline Chemical Actuators. *Chem. Mater.*, 15, 2411–2418, 2003, doi: 10.1021/cm020329e.

124. Okuzaki, H. and Kunugi, T., Adsorption-induced bending of polypyrrole films and its application to a chemomechanical rotor. *J. Polym. Sci. Part B Polym. Phys.*, 34, 1747–1749, 1996, doi: 10.1002/(SICI)1099-0488(19960730)34:10<1747::AID-POLB5>3.0.CO;2-N.

125. Okuzaki, H. and Kunugi, T., Adsorption-induced chemomechanical behavior of polypyrrole films. *J. Appl. Polym. Sci.*, 64, 383–388, 1997, doi: 10.1002/(SICI)1097-4628(19970411)64:2<383::AID-APP20>3.0.CO;2-0.

126. Conway, N.J., Traina, Z.J., Kim, S.-G., A strain amplifying piezoelectric MEMS actuator. *J. Micromech. Microeng.*, 17, 781–787, 2007, doi: 10.1088/0960-1317/17/4/015.

127. Detsi, E., Onck, P., De Hosson, J.T.M., Metallic Muscles at Work: High Rate Actuation in Nanoporous Gold/Polyaniline Composites. *ACS Nano*, 7, 4299–4306, 2013, doi: 10.1021/nn400803x.

128. Li, D., Wang, Y., Xia, Y., Electrospinning of Polymeric and Ceramic Nanofibers as Uniaxially Aligned Arrays. *Nano Lett.*, 3, 1167–1171, 2003, doi: 10.1021/nl0344256.

129. Teo, W.E. and Ramakrishna, S., Electrospun fibre bundle made of aligned nanofibres over two fixed points. *Nanotechnology*, 16, 1878–1884, 2005, doi: 10.1088/0957-4484/16/9/077.

130. Shin, M.K., Kim, S.I., Kim, S.J., Kim, S.-K., Lee, H., Spinks, G.M., Size-dependent elastic modulus of single electroactive polymer nanofibers. *Appl. Phys. Lett.*, 89, 231929, 2006, doi: 10.1063/1.2402941.

131. Spinks, G.M., Mottaghitalab, V., Bahrami-Samani, M., Whitten, P.G., Wallace, G.G., Carbon-Nanotube-Reinforced Polyaniline Fibers for High-Strength Artificial Muscles. *Adv. Mater.*, 18, 637–640, 2006, doi: 10.1002/adma.200502366.

132. Lu, W., Norris, I.D., Mattes, B.R., Electrochemical Actuator Devices Based on Polyaniline Yarns and Ionic Liquid Electrolytes. *Aust. J. Chem.*, 58, 263, 2005, doi: 10.1071/CH04255.

133. Bhardwaj, N. and Kundu, S.C., Electrospinning: A fascinating fiber fabrication technique. *Biotechnol. Adv.*, 28, 325–347, 2010, doi: 10.1016/j.biotechadv.2010.01.004.

134. Picciani, P.H.S., Medeiros, E.S., Orts, W.J., Mattoso, L.H.Z., Advances, L.H.C., Advances in Electroactive Electrospun Nanofibers in Electroactive, in: *Nanofibers - Production, Properties and Functional Applications*, T. Lin (ed). Intech Open, China, 85–116, 2011, ISBN-10: 9533074205.

8

Fluid Power Actuators

Mohanraj Thangamuthu[1], Tamilarasi Thangamuthu[2],
Gobinath Velu Kaliyannan[3], Moganapriya Chinnasamy[3]
and Rajasekar Rathinasamy[3]*

[1]*Department of Mechanical Engineering, Amrita School of Engineering, Amrita
Vishwa Vidyapeetham, Coimbatore, India*
[2]*Department of Mechatronics Engineering, Kongu Engineering College,
Perundurai, Erode, India*
[3]*Department of Mechanical Engineering, Kongu Engineering College, Perundurai,
Erode, India*

Abstract

Fluid power is the use of fluids (liquid/air) under high pressure to generate, control, and transmit power. Fluid power systems are used to transmit power from a central source to industrial users over extended geographic areas. Fluid power actuator consists of cylinder or motor that uses fluid power to assist mechanical operation. The mechanical motion gives an output in terms of linear, rotatory, or oscillatory motion. This chapter deals the different types of fluid power actuators and its application in the field of mechanical engineering.

Keywords: Fluid power, hydraulics, pneumatics, actuators, cylinders, motors, valves

8.1 Introduction

An actuator is an energy converter that converts energy from an external source into mechanical energy in a convenient way. The type of actuator depends on the input parameters used in the energy conversion process. For example, the input quantities of electromagnetic, piezoelectric and fluid

---

Corresponding author: rajasekar.cr@gmail.com

Inamuddin, Rajender Boddula and Abdullah M. Asiri (eds.) Actuators: Fundamentals, Principles, Materials and Applications, (187–210) © 2020 Scrivener Publishing LLC

power actuators can be the current, charge, and fluid pressure. Making use of the actuators, one can perform the required task.

8.2   Classification of Actuators

Actuators are classified based on the material used as shown in Table 8.1 [1]. Fluid power actuators are one type of actuator which receives fluid either in the form of liquid or gas normally driven by an electric motor. This actuator transforms its energy into a linear or rotary motion to do a valuable task. Fluid power actuators are normally concerned with movement and application force on the object. Fluid power actuators are mainly classified into two types, i.e., hydraulic and pneumatic actuators. Pneumatic and hydraulic actuators are built in the same manner but differ in the operating pressure [2].

The amount of a liquid or air flow is corrected by means of appropriate control valves like directional, pressure, and flow. Electric input supply is used to control these devices and transform the applied power into a specific parameter like stroke, speed, and force of a longitudinal or rotational motion [3]. Further, fluid power actuators are classified into three types, namely, linear, rotary, and semi-rotary actuators.

8.2.1   Linear Actuator or Cylinders

A straight line or linear motion is provided by the linear actuators. The purpose of cylinder is to convert fluid power into linear mechanical movement. Extension and retraction of a piston in the cylinder provides a push or pull force to drive the external load along a straight line path.

Table 8.1 Materials used for actuators.

Type of actuator	Example
Fluid power	Hydraulics, Pneumatics
Pizoelectric	Ceramic, Polymer
Natural	Human muscle
Electro-mechanical	Linear drive, MEMS Comb Drives
Smart materials	Shape Memory Alloy, Bimetallic
Hybrid	Piezoelectric, Electro-mechanical
Electromagnetic	Solenoid, Magnetostriction

8.2.2 Rotary Actuator or Motors

Rotary actuators convert fluid energy into mechanical energy by providing rotary motion like clockwise or counterclockwise by the application of pressure similar to electrical drives. The speed of rotation can be controlled by regulating the flow rate of the fluid. However, the torque relies on variation in the motor inlet and outlet pressure. Continuous angular movements are obtained by a rotary actuator, known as a hydraulic motor. Hydraulic motors come with less speed/more torque, more speed/less torque, and a few are less or more speed/more torque.

8.2.3 Semi-Rotary Actuators

This type of actuators is competent in restricted angular movements which can be partial rotations or less than 360° is more normal.

8.3 Hydraulic Actuator

The hydraulic system is utilized to generate, control, and transmit power through pressurized liquid [4]. The basic components involved to drive the actuator in a hydraulic system is shown in Figure 8.1 [2]. A pump is used for driving an electric motor to make fluid flow, where the pressure, direction, and fluid flow rate are varied through a different set of valves.

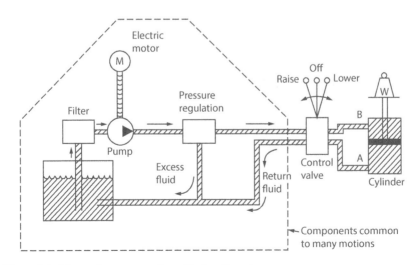

Figure 8.1 Schematic representation of hydraulic system.

Hydraulic actuators are used for converting hydraulic energy into mechanical energy. They are also employed for controlling and transmitting the power up to 420 bar with lesser velocities. The output power depends on the rate of flow, pressure drop, and overall efficiency. The key feature of hydraulics is a larger power to weight ratio compared to pneumatics and electrical systems and appropriate for working in hard as well as robust environment [5].

8.4 Pneumatic Actuator

The pneumatic system uses a filter, regulator, and lubricator (FRL) unit to clean, regulate and lubricate the dry air generated from the atmosphere. Air from the FRL unit is given to the pneumatic actuator through direction and flow control valve. The pneumatic actuator normally used for lower pressure in the range of 6 to 10 bar with high velocities. So as to increase the accuracy of fluid power actuators, it is required to follow the system parameters and should compensate for the errors due to environmental disturbances. It is necessary to use the concept of closed-loop feedback to control the variable and constantly comparing it by means of the comparator with the standard values. Pneumatic actuators are appropriate to transport light weights with high velocities [3]. The basic schematic of the pneumatic system components was presented in Figure 8.2 [2]. Fluid power actuators are classified based on actuation types:

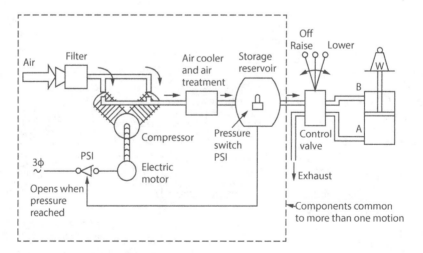

Figure 8.2 Schematic representation of pneumatic system.

- Single-acting cylinders
- Double-acting cylinders
- Telescopic cylinders
- Tandem cylinders

8.4.1   Single-Acting Cylinders

A single-acting cylinder (SAC) contains a piston surrounded by a cylindrical housing called barrel. The rod in the piston is connected with one end of the cylinder which can be used to produce reciprocating motion. Figure 8.3 shows the representation of single-acting cylinder. SACs are used to produce a force that may be extension or retraction in only one direction by fluid pressure applied on the piston. Because of the single port in SAC, the other stroke of the piston is not done with fluid pressure, and the retraction stroke is accomplished either by gravity or spring tension. While the SAC is mounted vertically up and retraction is done through the self-weight. When the SAC is mounted horizontally, the spring tension is used for retraction. Because of the return mechanism of the piston rod, SACs are categorized as:

- Gravity return SAC
- Spring return SAC

8.4.1.1   Gravity Return SAC

Figure 8.4 shows the push- and pull-type gravity returns SAC. In push type, the weight (W) is lifted against gravity by means of fluid pressure during the extension stoke. The fluid either gas or liquid is passed through the pressure port as shown in Figure 8.4a. The vent is kept open to the atmosphere so that air can easily enter in and leave from the cylinder.

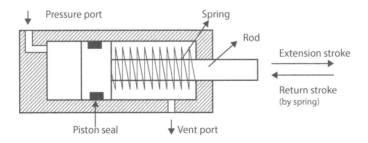

Figure 8.3 Single-acting cylinder.

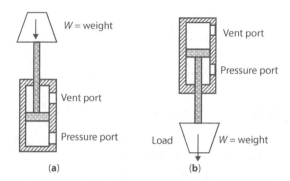

Figure 8.4 Gravity return SAC: (a) Push type; (b) Pull type.

During the retraction stroke of the cylinder, the pressurized fluid is removed from cylinder through the vent port. This permits the weight to move forward the fluid out from the cylinder to the reservoir. In pull-type gravity return SAC, the extension of the cylinder is carried out by means of weight added at the piston rod. This type of cylinder extends mechanically when the pressure port is connected with the tank.

8.4.1.2   SAC With Spring Return

SAC with spring return was shown in Figure 8.5. Similar to gravity return cylinder, push-type and pull-type spring return cylinders are normally used for extending and retracting the piston rod for the required application. In push-type spring return SAC, compressed air is passing through the pressure port, and the cylinder starts extends. The spring automatically retracts the cylinder to its home position (retraction) when the pressure port is released. The vent port is kept open to vent so that air can flow easily enter in and leave from the cylinder. In pull-type, normally, the cylinder is in a fully extended position only. When the pressurized fluid passes through the pressure port, the cylinder starts retracting. In this cylinder, the pressure port is located at the cylinder's rod end.

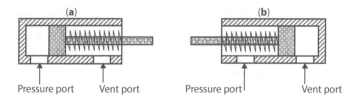

Figure 8.5 (a) Push and (b) pull type SAC.

8.4.2 Double-Acting Cylinder

Double-acting cylinder (DAC) exchanges the cycles of high-pressure fluid to both sides of the piston and produces extension and retraction forces to move the piston, allowing more control over the movement. There are two types of DAC:

- Double-acting cylinder with a piston rod on one side
- Double-acting cylinder with a piston rod on both sides

8.4.2.1 DAC With Piston Rod on One Side

Figure 8.6 illustrates the working of DAC with a piston rod on a single side. During the extension stroke of the cylinder, the pressurized fluid is sent through pressure port (pump flow) as shown in Figure 8.6a. Fluid from return flow (vent port) returns to the reservoir. While retraction of the cylinder, the pressurized fluid is sent to the vent port and fluid from pump flow returns to the tank as shown in Figure 8.6b.

8.4.2.2 DAC With Piston Rod on Both Sides

A DAC with a piston rod on both sides is shown in Figure 8.7. It is employed in an application where work can be done by two ends of the cylinder,

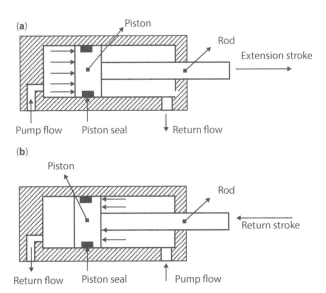

Figure 8.6 DAC with piston on one side (a) Extension stroke and (b) Return stroke.

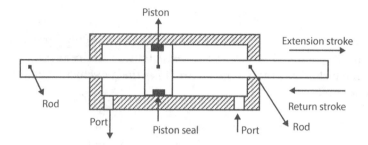

Figure 8.7 DAC with a piston rod on one side.

thus making it more useful. Double rod cylinders can endure elevated side loads as they contain extra bearing, one on each rod, to resist the loading.

8.5 Telescopic Cylinder

A telescopic cylinder is employed where larger extension stroke length and the shorter retraction stroke length is required. Figure 8.8 shows the three-stage telescopic cylinder. The extension of the telescopic cylinder has different phase, each phase has a sleeve which fits within the previous phase. A telescopic cylinder may be a single-acting and double-acting type. Due to their more complex construction, they are costlier than normal

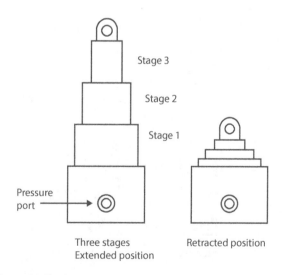

Figure 8.8 Telescopic cylinder.

cylinders. Normally, it consists of a shell of tubes and operates on the displacement principle. Tubes are grasped by means of bearing rings, inmost set has grooves to permit the fluid flow. The seals and wiper rings are used in the front side bearing assembly. Stop rings are used to restrict the movement of each section, thus prevents the partition.

When the telescopic cylinder starts to move forward, the whole unit moves together till the complete extension of exterior unit. The remaining units move forward until the second exterior unit reaches the end of its extension. The same course of action is repeated until all the units are fully extended. The speed of operation increases for a given input flow rates in steps as each succeeding unit reaches its full extension. Intended for a particular value of pressure, the load-carrying capacity of the cylinder falls down for every subsequent unit. Single-acting telescopic cylinders are typically climbed perpendicularly with the smaller ram. The cylinder may weight returned due to the up and down movement of the ram. It also provides the large ram with its ports attached to a stationary machine member.

8.6   Tandem Cylinder

Tandem cylinders used in applications can generate nearly double the force from the same diameter of the normal cylinder. Figure 8.9 shows the tandem cylinder. Pressurized air is enforced to both ends of pistons, resulting in augmented force. The problem in this type of cylinder is its larger length.

8.7   Research Towards the Applications of Pneumatic Fluid Power Actuators

The pneumatic actuator is mainly used for a force control system in numerous engineering applications. The static and dynamic characteristics of the

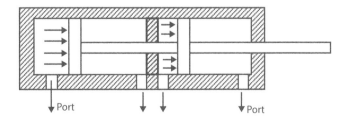

Figure 8.9 Tandem cylinder.

fluid power actuator play a vital role in the overall performance of the control system. Hence, enhancing the dynamic performance of actuator is the main task for the control system designers. Many control approaches are deployed on the pneumatic actuators in terms of modeling, simulation, and control methods applied for various applications. Pneumatic systems are widely used in automatic controllers and production applications. Several types of pneumatic controllers are used for improving the performances like fuel consumption, dynamic response, and output stiffness [6].

By using normal compliance, pneumatic actuators might imitate natural spring and damping characteristics in equipment like chair [7], robotic forceps [8], ball and beam system [9], rehabilitation robotics [10, 11], robotic leg [12], exoskeleton [13], and more. Even though pneumatic systems are very sensitive to small changes in air pressure or flow rate, safe properties of pneumatic systems are appropriate for human-friendly application. The majority of soft robotics [14, 15] apply pneumatic sources to drive the system like medical applications [16], imitate biological creatures [17–19], Giacometti structures [20, 21], biomimetic hands [22, 23], and others.

The pneumatic servo system has more advantages than the hydraulic system in elevated temperature and nuclear environments. The pneumatic actuators have numerous advantages like easy handling, maintenance, minimal cost, safe, and easy to install. The pneumatic actuator instead of the servo valve normally limits system response and rigidity. The cylinder piston is possibly the most excellent alternative where simplicity and cost are dominant. But the rotary type of motor is directed if minimum fuel consumption is desired [24].

Singh et al. depicted the design for air brake control valves for heavy as well as medium-sized trucks. For this application, a floor-mounted pneumatic valve working on advanced design was formulated with electrical controls. The performance of the system was verified on the truck and experienced with appropriate system requirements of federal brake regulations [25]. The pneumatic actuator was used to precisely control the joint position of the intelligent soft arm control (ISAC) robot system. Initially, a physical model of the actuator was considered and used as the base for auxiliary torque control. The developed model was established by implementing it as a torque controller and tested by conducting the experiments [26].

Pneumatic actuators are widely employed for transporting granular materials from one place to another. A method that combines the theoretical as well as experimental methods to characterize and calculate the friability of granules in the laboratory-scale system was designed [27]. Kim et al. (2008) formulated a well-organized robotic deburring approach based on a novel pneumatic actuator. The developed approach has measured the interaction between manipulator and workpiece. The dynamics and control design of the tool was clearly believed in the process of deburring. The novel dynamic pneumatic tool was designed based on a single pneumatic actuator along with the inactive chamber for delivering the conformity and decrease the chatter originated due to the compressible nature of air [28].

A coordination control method was deployed, that implements two levels of the hierarchical control structure. The unwanted influence of surrounding turbulence like friction and nonlinear conformity of pneumatic cylinder slowing from the compressible nature of air was minimized by the method of feedback linearization. The coordination control method illustrated its effectiveness in terms of deburring accuracy and speed [28].

8.7.1 Novel Micro-Pneumatic Actuator for MEMS

Reduction of weight, size, and power utilization is the major issue in micro-technology and bioengineering, which requires the accessibility of smart micro-mechanical actuators. The compressed air is used as a driving force in the micro-pneumatically driven actuator. This type of actuator has several merits:

- high design flexibility
- large displacements
- higher forces
- outstanding dynamic performance
- high energy density
- utilization of different fluids as driving medium

The performance of actuators and grasping forces produced by the pneumatically operated microgripper was analyzed by employing a unique setup that includes a force sensor. The micro-pneumatic actuator includes a piston that is associated with the housing by two springs. These springs facilitate the cylinder piston to move while pressure was applied and also act as a seal against the environment. Two Pyrex wafers were developed on the top and bottom of the cylinder. Figure 8.10 shows the structure of the microgripper [29].

The linear movement of piston and spring was enabled by a small distance between Pyrex lids and fasten structure of microns. During the process of wafer dicing, the cylinder inlet was opened. The piston was connected to the gripper linkage by a hinge for transmitting the produced

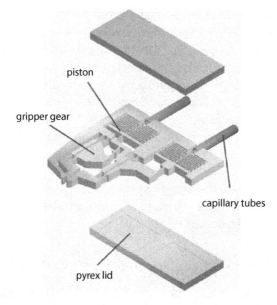

piston

gripper gear

capillary tubes

pyrex lid

Figure 8.10 Structure of pneumatically driven micro gripper.

force. The hinge reimburses the rotation of the gripper. Two pneumatic actuators were used for this application one for opening, and one for closing the microgripper.

Normally, two types of fabrication processes are used in this process. They are silicon RIE and UV-depth lithography with SU8. Long-term stability and no fatigue characteristics of silicon are used as the base material. Additional merit is anodic bond technology and the prospect of incorporating sensors based on diffused piezo-resistors. The SU8 offers good portability of accommodating structures owing to the lesser modulus of elasticity. The SEM image of the silicon structure was shown in Figure 8.11. To enhance the assembling accuracy, visual inspection is repeatedly necessary. To assess the actuator's performances, a test rig was constructed and tested. It has a tactile 3D force sensor and a high precision XYZ stage. It is used to enable an accurate positioning against the force sensor. The force sensor can be located with the deflection facility in an accurate manner. The measurements were performed with the application of constant displacement through the force sensor to a pneumatic actuator. The characteristic curve of the developed pneumatic actuator with various values of pressure was presented in Figure 8.12.

The pressure of about 120 mbar displacement is obtained and produced forces of over 10 mN were obtained with the silicon piston bellow

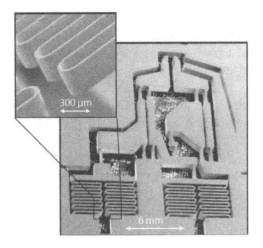

Figure 8.11 SEM image of silicon structure.

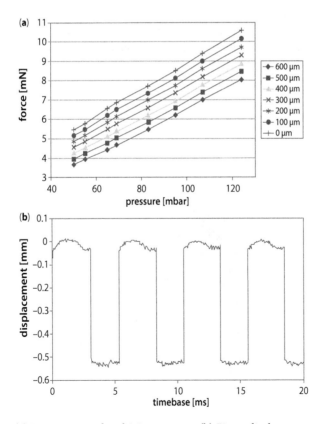

Figure 8.12 (a) Force generated vs driving pressure; (b) Piston displacement.

(Figure 8.13a). The frequency of 130 Hz and strokes of 300 μm was measured (Figure 8.13b). The produced gripper force was also measured with the same test and maximum grasping forces up to 10 mN were reached with a pressure of 400 mbar.

A robot with an artificial musculoskeletal system for biomechanics research was proposed [30]. The pneumatic artificial muscle was used in the Bipedal robot. The results revealed that the robot can perform vertical jumping, soft landing from a 1-m height, and postural control during standing. The Bipedal robot is structurally similar to an animal and used for biomechanics research. This musculoskeletal robot helps to enlighten the design principles of the robot which can move rapidly and dexterously in the real world. Figure 8.14 shows the Bipedal robot with an artificial musculoskeletal system developed using electro-pneumatic controls [30].

A jumping robot with an anthropomorphic muscular skeleton structure was developed using pneumatic artificial muscles. The developed jumping robot with pneumatic artificial muscles can mimic the structure and function of a human leg [31]. A detailed review on applications of Pneumatic Artificial Muscles (PAMs) was presented. PAMs have the merits like high strength and small weight attracts more applications in industries as well as scientific utilization [32].

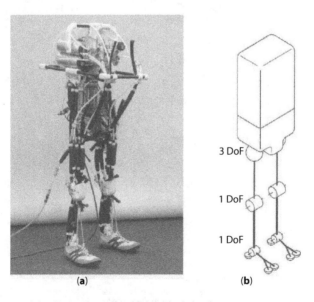

(a)   **(b)**

Figure 8.13 Bipedal robot with artificial musculoskeletal system.

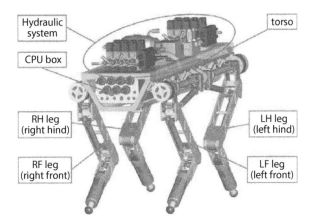

Figure 8.14 CAD model of HyQ Leg with on-board system.

8.8   Research Towards the Applications of Hydraulic Fluid Power Actuators

Currently, digital fluid power techniques have increasingly entered into industries. The digital hydraulic fluid power system has numerous merits like precision, maximum efficiency, robustness, and reliability compared to other technologies. It provides new functionality such as sensor less incremental actuation and digital control for multiple parts arrangement that is difficult with present fluid power systems. It is confident in creating the digital system with salient features and also affords new prospects for fluid power engineering in the future [33].

The first hydraulic legged robot (BigDog) was designed and developed by Boston Dynamics™ using hydraulic actuators in 2005 [5]. BigDog was the first advanced rough hydraulic robot to leave the lab and adjust to the real-world environment. BigDog was developed with hydraulic actuators to accompany soldiers in rough terrain which is difficult to move with normal vehicles [34]. The development of various hydraulic legged robots was shown in Table 8.2.

Hydraulic Quadruped (HyQ) is a multipurpose hydraulically powered quadruped robot, used to study the dynamic movements as well as navigation performance. Figure 8.14 illustrates the CAD model of the HyQ Leg with the onboard system [35]. In 1990s, John Deere's R&D unit in Finland designed a prototype of a harvester. John Deere forestry walking machine was shown in Figure 8.15 which mainly used for the logging

Table 8.2 Details of hydraulic legged robot.

Name	Height (m)	Weight (kg)	Payload (kg)	Power (kW)	Actuation	Joint
GE truck [39]	3	1,400	–	7.46	Hydraulic	12
ASV [40]	2	–	–	52.2 Gasoline engine	Hydraulic	18
Raibert's monopod [41]	1.1	17.3	–	–	Hydraulic	3
Raibert's quadruped [41]	0.35	25.2	–	–	Hydraulic	8
BogDog [34, 42]	1	109	45	Gasoline	Hydraulic	16
TITAN XI [43]	2	7,000	–	2.8 lt engine	Hydraulic	12
ROBOCLIMBER [44]	0.65	3,500	4,000	Hydraulic	Hydraulic	12
HyQ [35]	0.5	91	–	AC source	Hydraulic	16
COMET-IV [45]	3.3	2,120	–	Gasoline	Hydraulic	16
SCalf-1 [46]	0.4	65	80	–	Hydraulic	12
Quadruped robot [47]	–	–	50	Battery	Hydraulic	16
Baby elephant [48]	0.5	130	60	Battery	Motor & hydraulic	12
Wildcat [49]	1.17	154	30	Battery	Hydraulic	14
LS3 [50]	1.7	590	181	Gas and diesel	Hydraulic	12
RL-A1 [51]	1	60.2	120	–	Hydraulic	16
Jinpoong [52]	1.2	120	–	2 power pack engine	Hydraulic	16
MiniHyQ [53]	0.77	35	–	AC source	Hydraulic	12
Spot [54]	0.94	75	45	Battery	Hydraulic	12
Handle [55]	2	105	45	Battery	Hydraulic & electric	10
HyQ2Max [46, 56]	0.54	80	40	AC source	Hydraulic	12

process. This new machine offers perception into new knowledge which is being developed and probably change the future logging industry and setting a path for further progression in productive and environment-friendly machinery [4].

Figure 8.15 John Deere forestry walking machine [4].

The quadruped walking robot qRT-2 is a front-drive instrument with fluid power actuators and wheeled back legs, actuated by a hydraulic system developed by the Korea Institute of Industrial Technology as presented in Figure 8.16a [36]. The weight of qRT-2 is around 60 kg with a payload of 40 kg. By this method, the gait runs at a speed of 1.3 m/s on a smooth surface and walking at 0.7 m/s on a rough surface. The team also developed a four-leg walking robot P2 and studied the hydraulic flow consumption while robot in walking stage [37]. For minimization of the hydraulic flow consumption, a power optimization scheme was deployed in P2. The simulation and experimental results demonstrated the potential performance. All joints of the robotic legs were powered by tiny hydraulic actuators, as presented in Figure 8.16b.

Hydraulics is employed in various applications like huge and heavy industries and modern robotics. Due to the advantage of high power-to-weight (P/W) ratio compared that in another system, it makes appropriate for working in rough terrain. The innovation of hydraulic actuators and components were showed at smallness, reduced weight with enhanced efficiency, clear, and friendly features for various applications including mobile and industrial hydraulics. The hydraulic developments in rough robotics applications that focus on legged robots like humanoid, large machinery, and search-and-rescue robot were shown in Figure 8.17 [5].

The rapid expansion of the microelectromechanical system (MEMS) actuator has a new evolution in terms of efficiency, power, and force. The progress of helium balloon membrane actuators to piston-cylinder was described, and drag based micro-device was shown in Figure 8.18.

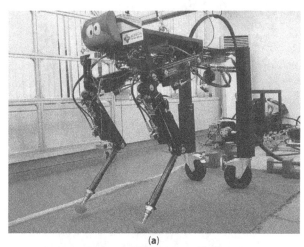

(a)

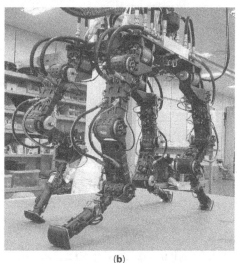

(b)

Figure 8.16 (a) Quadruped walking robot qRT-2; (b) Rotary actuators used in walking robot.

Elastic-type actuators expand under an applied pressure which consists of any type of membrane. These types of actuators have no sliding parts due to the advantage of less friction, wear, and sealing issues. The various executions of elastic actuators materialized over the years, categorized into membrane, bellows, balloon, and artificial muscle actuators as shown in Figure 8.19.

De Volder and Reynaerts afford an outline and categorization of actuators based on their principle of actuation. Generally, used piston-cylinder–based

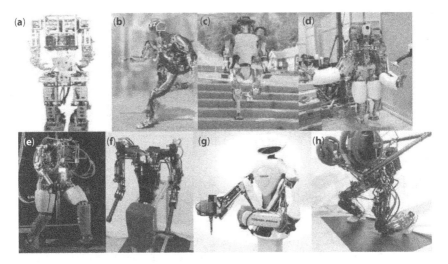

Figure 8.17 Hydraulic humanoid robots [5].

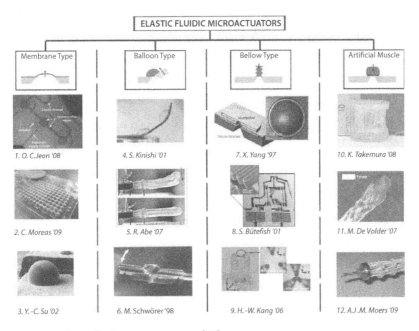

Figure 8.18 Elastic fluidic micro actuators [38].

microactuators are able to attain larger strokes as well as forces. Current trends show gradually more hydraulic systems that include integrated pumps, valves, sensors, and actuator arrays. Compared with other principles of actuation, a fluid power system provides remarkable advantages in

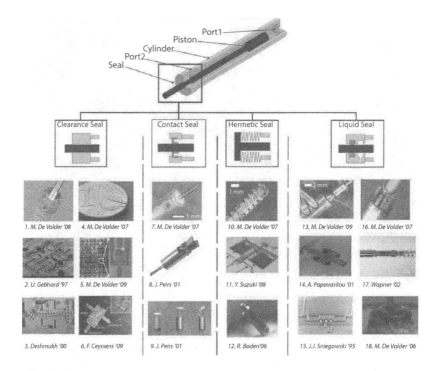

Figure 8.19 Classification of elastic actuators [38].

terms of low cost and higher force. The major field of application includes μTAS, apparatus for simple invasive surgery, as well as micro robotics [38].

References

1. Zupan, M., Ashby, M.F., Fleck, N.A., Actuator classification and selection— The development of a database. *Adv. Eng. Mater.*, 4, 933–940, 2002.
2. Parr, A., *Hydraulics and pneumatics: A technician's and engineer's guide*, Elsevier, UK, 2011.
3. Janocha, H., *Actuators*, Springer, Berlin, 2004.
4. Yang, H.-y. and Pan, M., Engineering research in fluid power: A review. *J. Zhejiang Univ.-Sci. A*, 16, 427–442, 2015.
5. Suzumori, K. and Faudzi, A.A., Trends in hydraulic actuators and components in legged and tough robots: A review. *Adv. Rob.*, 32, 458–476, 2018.
6. Ali, H.I. *et al.*, A review of pneumatic actuators (modeling and control). *Aust. J. Basic Appl. Sci.*, 3, 440–454, 2009.
7. Faudzi, A.A.M., Suzumori, K., Wakimoto, S., Design and control of new intelligent pneumatic cylinder for intelligent chair tool application, in: *2009*

IEEE/ASME International Conference on Advanced Intelligent Mechatronics, IEEE, 2009.

8. Haraguchi, D., Tadano, K., Kawashima, K., Development of a pneumatically-driven robotic forceps with a flexible wrist joint. *Procedia CIRP*, 5, 61–65, 2013.

9. Azman, M.A. *et al.*, Integrating servo-pneumatic actuator with ball beam system based on intelligent position control. *J. Teknol.*, 69, 73–79, 2014.

10. Osman, K., Rahmat, M., Suzumori, K., Intelligent pneumatic assisted therapy on ankle rehabilitation, in: *2015 IEEE International Conference on Rehabilitation Robotics (ICORR)*, IEEE, 2015.

11. Saga, N., Saito, N., Nagase, J.-y., Ankle Rehabilitation Device to Prevent Contracture Using a Pneumatic Balloon Actuator. *IJAT*, 5, 538–543, 2011.

12. Colbrunn, R.W., Nelson, G.M., Quinn, R.D., Design and control of a robotic leg with braided pneumatic actuators, in: *Proceedings 2001 IEEE/RSJ International Conference on Intelligent Robots and Systems. Expanding the Societal Role of Robotics in the the Next Millennium (Cat. No. 01CH37180)*, IEEE, 2001.

13. Hyon, S.-H. *et al.*, Design of hybrid drive exoskeleton robot XoR2, in: *2013 IEEE/RSJ International Conference on Intelligent Robots and Systems*, IEEE, 2013.

14. Rus, D. and Tolley, M.T., Design, fabrication and control of soft robots. *Nature*, 521, 467, 2015.

15. Majidi, C., Soft robotics: A perspective—Current trends and prospects for the future. *Soft Rob.*, 1, 5–11, 2014.

16. Ranzani, T. *et al.*, A bioinspired soft manipulator for minimally invasive surgery. *Bioinspiration Biomimetics*, 10, 035008, 2015.

17. Niiyama, R., Nagakubo, A., Kuniyoshi, Y., Mowgli: A bipedal jumping and landing robot with an artificial musculoskeletal system, in: *Proceedings 2007 IEEE International Conference on Robotics and Automation*, IEEE, 2007.

18. Kingsley, D.A., Quinn, R.D., Ritzmann, R.E., A cockroach inspired robot with artificial muscles, in: *2006 IEEE/RSJ International Conference on Intelligent Robots and Systems*, IEEE, 2006.

19. Razif, M.R.M. *et al.*, Two chambers soft actuator realizing robotic gymnotiform swimmers fin, in: *2014 IEEE International Conference on Robotics and Biomimetics (ROBIO 2014)*, IEEE, 2014.

20. Takeichi, M. *et al.*, Development of Giacometti arm with balloon body. *IEEE Rob. Autom. Lett.*, 2, 951–957, 2017.

21. Faudzi, A.A.M. *et al.*, Long-legged hexapod Giacometti robot using thin soft McKibben actuator. *IEEE Rob. Autom. Lett.*, 3, 100–107, 2017.

22. Wang, Z., Torigoe, Y., Hirai, S., A prestressed soft gripper: Design, modeling, fabrication, and tests for food handling. *IEEE Rob. Autom. Lett.*, 2, 1909–1916, 2017.

23. Faudzi, A.A.M. *et al.*, Index finger of a human-like robotic hand using thin soft muscles. *IEEE Rob. Autom. Lett.*, 3, 92–99, 2017.

24. Tablin, L. and Gregory, A., Rotary pneumatic actuators. *J. Control Eng.*, 58–63, 1963.
25. Singh, H., Lang, P.R., Auman, J.T., Centralized electro-pneumatic control system for truck air brakes. *SAE Trans.*, 94, 1127–1134, 1985.
26. Schroder, J. *et al.*, Dynamic pneumatic actuator model for a model-based torque controller, in: *Proceedings 2003 IEEE International Symposium on Computational Intelligence in Robotics and Automation. Computational Intelligence in Robotics and Automation for the New Millennium (Cat. No. 03EX694)*, IEEE, 2003.
27. Rajniak, P. *et al.*, Modeling and measurement of granule attrition during pneumatic conveying in a laboratory scale system. *Powder Technol.*, 185, 202–210, 2008.
28. Kim, C., Chung, J.H., Hong, D., Coordination control of an active pneumatic deburring tool. *Rob. Comput.-Integr. Manuf.*, 24, 462–471, 2008.
29. Bütefisch, S., Seidemann, V., Büttgenbach, S., Novel micro-pneumatic actuator for MEMS. *Sens. Actuators A: Phys.*, 97, 638–645, 2002.
30. Niiyama, R. and Kuniyoshi, Y., Pneumatic biped with an artificial musculoskeletal system, in: *Proceedings of 4th International Symposium on Adaptive Motion of Animals and Machines*, 2008.
31. Hosoda, K. *et al.*, Pneumatic-driven jumping robot with anthropomorphic muscular skeleton structure. *Auton. Robots*, 28, 307–316, 2010.
32. Andrikopoulos, G., Nikolakopoulos, G., Manesis, S., A Survey on applications of Pneumatic Artificial Muscles, in: *2011 19th Mediterranean Conference on Control & Automation (MED)*, 2011.
33. Scheidl, R., Linjama, M., Schmidt, S., Is the future of fluid power digital? *Proceedings of the Institution of Mechanical Engineers, Part I: Journal of Systems and Control Engineering*, vol. 226, pp. 721–723, 2012.
34. Raibert, M. *et al.*, Bigdog, the rough-terrain quadruped robot. *IFAC Proceedings Volumes*, vol. 41, pp. 10822–10825, 2008.
35. Semini, C. *et al.*, Design of HyQ–a hydraulically and electrically actuated quadruped robot. *Proceedings of the Institution of Mechanical Engineers, Part I: Journal of Systems and Control Engineering*, vol. 225, pp. 831–849, 2011.
36. Kim, H. *et al.*, Foot Trajectory Generation of Hydraulic Quadruped Robots on Uneven Terrain. *IFAC Proceedings Volumes*, vol. 41, pp. 3021–3026, 2008.
37. Kim, T.-J. *et al.*, The energy minimization algorithm using foot rotation for hydraulic actuated quadruped walking robot with redundancy, in: *ISR 2010 (41st International Symposium on Robotics) and ROBOTIK 2010 (6th German Conference on Robotics)*, VDE, 2010.
38. De Volder, M. and Reynaerts, D., Pneumatic and hydraulic microactuators: A review. *J. Micromech. Microeng.*, 20, 043001, 2010.
39. Liston, R. and Mosher, R., A versatile walking truck, in: *Transportation Engineering Conference*, 1968.

40. Pugh, D.R. *et al.*, Technical description of the adaptive suspension vehicle. *Int. J. Rob. Res.*, 9, 24–42, 1990.
41. Raibert, M.H., *Legged robots that balance*, MIT press, England, 1986.
42. Playter, R., Buehler, M., Raibert, M., BigDog, in: *Unmanned Systems Technology VIII*, International Society for Optics and Photonics, 2006.
43. Hodoshima, R. *et al.*, Development of TITAN XI: A quadruped walking robot to work on slopes, in: *2004 IEEE/RSJ International Conference on Intelligent Robots and Systems (IROS)(IEEE Cat. No. 04CH37566)*, IEEE, 2004.
44. Nabulsi, S. *et al.*, High-resolution indirect feet–ground interaction measurement for hydraulic-legged robots. *IEEE Trans. Instrum. Meas.*, 58, 3396–3404, 2009.
45. Irawan, A. and Nonami, K., Compliant walking control for hydraulic driven hexapod robot on rough terrain. *J. Rob. Mechatronics*, 23, 149–162, 2011.
46. Rong, X. *et al.*, Design and simulation for a hydraulic actuated quadruped robot. *J. Mech. Sci. Technol.*, 26, 1171–1177, 2012.
47. RunBin, C. *et al.*, Inverse kinematics of a new quadruped robot control method. *Int. J. Adv. Rob. Syst.*, 10, 46, 2013.
48. Wang, J., Gao, F., Zhang, Y., High power density drive system of a novel hydraulic quadruped robot, in: *ASME 2014 International Design Engineering Technical Conferences and Computers and Information in Engineering Conference*, American Society of Mechanical Engineers, 2014.
49. Robots, L., *Boston Dynamics Official Website*, Wildcat, United States of America, 2019, Available from: https://www.bostondynamics.com/legacy.
50. Robots, L., *Boston Dynamics Official Website*, LS3, United States of America, 2019, Available from: https://www.bostondynamics.com/legacy.
51. Kawabata, K. *et al.*, Development of hydraulic quadruped walking robot "RL-A1", in: *Proc. Robot. Mechatronics Conf.*, 2014.
52. Cho, J., Park, S., Kim, K., Design of mechanical stiffness switch for hydraulic quadruped robot legs inspired by equine distal forelimb. *Electron. Lett.*, 51, 33–35, 2014.
53. Khan, H. *et al.*, Development of the lightweight hydraulic quadruped robot—MiniHyQ, in: *2015 IEEE International Conference on Technologies for Practical Robot Applications (TePRA)*, IEEE, 2015.
54. ROBOTS, *Boston Dynamics Official Website*, Spot, 2019, Available from: https://www.bostondynamics.com/spot.
55. ROBOTS, *Boston Dynamics Official Website*, HANDLE, United States of America, 2019, Available from: https://www.bostondynamics.com/handle.
56. Semini, C. *et al.*, Design overview of the hydraulic quadruped robots, in: *The Fourteenth Scandinavian International Conference on Fluid Power*, 2015.

9

Conducting Polymer/Hydrogel Systems as Soft Actuators

Yahya A. Ismail*, A. K Shabeeba, Madari Palliyalil Sidheekha and Lijin Rajan

Department of Chemistry, University of Calicut, Kerala, India

Abstract

Conducting polymers have received much attention in the field of actuators, but limits their applicability due to their low mechanical strength and processability. Conducting polymer/hydrogels (CPHs) are gels containing conducting polymer in the porous polymer hydrogel network possessing the combined properties of hydrogel and conductive system. Conducting polymer/hydrogel systems with decent electrical properties is attracting candidates for actuator applications due to high mechanical strength, biocompatibility, redox properties, stability, processability, etc. This chapter presents an overview of conducting polymer/hydrogel systems in the field of soft actuators. This chapter discusses a brief description of conducting polymer (CP) actuators with emphasis on the mechanism, merits and demerits. The purpose of this review is to summarize the progress of CP/hydrogel actuators fabricated using polypyrrole, polyaniline and polythiophene as the conducting moiety with various hydrogels, also outline the factors affecting their actuation performance and address future aspects.

Keywords: Conducting polymer, hydrogel, conducting polymer/hydrogels, actuators, polypyrrole, polyaniline, polythiophene, PEDOT:PSS

9.1 Introduction

Material science has been developed in each era according to the need and convenience of mankind, reached up to the era of plastics in last decades, and expected by the scientific community is that one of the future trends in material developing science will be the responsive soft materials including

Corresponding author: aiyahya@uoc.ac.in

Inamuddin, Rajender Boddula and Abdullah M. Asiri (eds.) Actuators: Fundamentals, Principles, Materials and Applications, (211–252) © 2020 Scrivener Publishing LLC

electroactive polymers (EAPs) and conductive hydrogels. Conducting polymer/hydrogels (CPHs) are gels containing conducting polymer (CP) in the porous polymer hydrogel network possessing the combined properties of hydrogel and conductive system. They have been received immense considerations due to their biocompatibility, safe nature, hydrophilic nature, mechanical strength, redox properties, etc.

Life and biological materials constitute a permanent inspiration for scientists and engineers to develop new biomimetic devices such as actuators. A material that can be termed as an actuator is the one which responds to an electric field, pH, temperature, ionic strength, and/or light with a corresponding change in their size, shape, and volume, that can perform mechanical work on a nano-, micro-, and macro scales [1–3]. By developing smart intelligent polymeric materials, we can make use of the effect of external parameters to adapt material combinations to get an auto control of a process with an actuator. Various types of materials like dielectric elastomers, shape memory alloys, carbon nanotubes, CPs, ionic polymer-metal composite, polymer gels, etc., have been proposed by the scientific community for the purpose of developing actuators [4]. CPH systems are of special interest as their electromechanical/electrochemical responses produce large strain when subjected to electrical stimulation and chemomechanical responses in some hydrogel systems. These properties make them efficient candidates for actuators that closely emulate natural muscles [5].

The applications of CPHs are not much explored because the design and fabrication of these systems are still a challenge. Owing to the emergence of reviews that discuss CPHs, this chapter presents an overview of CPH systems in the field of soft actuators that are not much reviewed. This chapter discusses a brief description of CP actuators and the fabrication of CPHs. This chapter covers the progress of CP/hydrogel actuators fabricated using polypyrrole, polyaniline, and polythiophene as the conducting moiety with various hydrogels and also outline the factors affecting their actuation performance.

9.2 Conducting Polymers as Actuators: A Brief Description

CPs are carbon-based polymers capable of mimicking biological functions and are electronically active [6, 7]. CPs have received much attention from both academic and industrial communities when Heeger, MacDiarmid, and Shirakawa were jointly awarded the Nobel Prize in Chemistry for their pioneering work on CPs in 2000. The ability of CPs to conduct electricity relies on the presence of conjugated π-bonds on their backbone, which helps in the migration of electrons throughout the chain [8]. CPs have received great interest owing to their outstanding features such as low cost, facile synthesis, controlled conductivity, lightweight, good stability, and intriguing electronic and redox properties [9–13], which are useful in most practical applications like actuators [14–19], chemical or biological sensors [20–23], electromagnetic shielding [24–27], electrochromic windows [28, 29], smart membranes [30–33], corrosion inhibitors [34, 35], solar cells [36–41], light-emitting diodes [42–45], batteries [46–50], super capacitors [51–54], etc. A number of reviews are available on these applications [55–61]. Un-doped CPs are insulators and become conducting or semiconducting when doped with suitable ions. The most extensively studied CPs are polyaniline (PANi), polypyrrole (PPy), polythiophenes (PTh), and their derivatives.

In recent decades, CPs have gained a reputable status from their use as actuators due to their biocompatibility, moderately high actuation strain, low operational voltage, and ability to generate forces similar to natural muscle [62, 63]. The first CP synthesized was polyacetylene by Shirakawa and coworkers in 1970 [64–66] and the first CP actuator date back to 1991 by Baughman *et al.* The actuators fabricated from CPs works in a similar way to that of natural muscles [5]. Recently, it has been reported that, due to the electrochemical nature of actuation, actuators developed from CPs can sense their working conditions such as applied current, temperature, electrolyte concentration, etc. while working [67–69].

9.2.1   Mechanism of CP Actuation

The redox process that takes place in the polymer chain of CP will be considered as the backbone of the actuation mechanism. The volume variation will motivate the linear and bending movement of the free-standing film, sheet, and fiber. Otero *et al.* had reported the concept of actuation [70]. The simple electrochemical reaction can explain the basics of electrochemical actuation in CP [69–72]. During oxidation, an electron is donated by the chain to the solvent and is called p-doping [73, 74]. The positive charge and free volume created will be compensated by the anion from the electrolyte and results in an increase in volume. It is an anion-driven oxidation (prevailing anion exchange). Reverse processes and a volume decrease is observed during reduction [69] (Figure 9.1a). If the reaction takes place in the presence of larger/macro anions like dodecylbenzene sulfonic acid (DBS), this anion remains trapped inside the polymer matrix. Therefore, the concentration of charge does not compensated by the anion exchange.

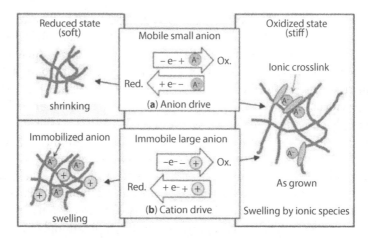

Figure 9.1 Schematic diagram of swelling deswelling (electrochemomechanical) mechanism. (a) shrinking of conducting polymer by reduction using small anion (prevailing anion exchange). (b) Expansion of conducting polymer by reduction using large anion (prevailing cation exchange). Reprinted with the permission from [78], copyright of Springer.

Here, during oxidation, the cation will expel out to balance the charge. The material shrinks during oxidation and swells during reduction (Figure 9.1b). Some other neutral polymers like PEDOT [75, 76], polythiophene [77], etc., can reduce within the medium. They have a high electronic affinity to get reduced [78] and attain a negative charge. The chain will swell with reduction and shrinks in the opposite action called n-doping.

The volume variation by the external voltage will force the system to attain linear or angular movement. The bending actuation can be demonstrated with a bilayer [70, 79, 80] or trilayer actuator, fabricated by CP. Bilayer actuator was the first developed actuator, reported by Otero *et al.* in 1992. They demonstrated a non-conducting tape bound with CP film generated by electrodeposition on a metallic electrode. By the electrochemical process within the system, mechanical stress is produced at the bilayer-tape interface. During this redox process, swelling and shrinking will drive the macroscopic movement of the bilayer free end (Figure 9.2). The CP will swell by the prevailing anion exchange and push the film into the convex side of the bending film. The opposite action will push the film into the concave side.

The metallic counter electrode has excluded from the three-layer actuator [80]. Here, two CP films were pasted on the two sides of double-sided tape (Figure 9.3), as reported by Otero *et al.* in 1992. During the electrochemical reaction in the electrolyte, one side of the film undergoes

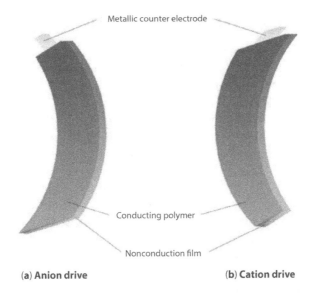

(a) Anion drive   **(b) Cation drive**

Figure 9.2 Schematic representation of bending of the bilayer actuator of conducting polymer. (a) Swelling by prevailing anion exchange on the concave side. (b) Shrinking by cation drive reduction on the convex side. Reprinted with the permission from [81], copyright of MDPI.

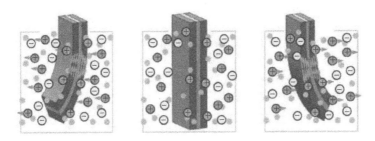

Figure 9.3 Schematic representation of the bending of the three-layer actuator using a conducting polymer. Swells by the oxidation and shrink by the reduction. Reprinted with the permission from [84], copyright of Royal Society of Chemistry.

swelling by anion exchange and another side of the film undergoes cation driven deswelling [82, 83]. The three-layer actuator can move outside the electrolyte by using a solid electrolyte formed by the ionic conducting membrane.

A linear movement is also possible in CP actuators by the expansion and contraction of the system in three dimensions by the insertion and de-insertion of the ions from the electrolyte [85].

9.2.2   Merits and Demerits of CP Actuators

The studies on CP actuators are still on progress and it reveals that these systems have potential possibilities, even though some limitations have to be overcomed before used in practical applications. Mainly, CP actuators are developed in view of developing devices whose performance would be similar to that of natural muscle. It is essential to have fast response time, high efficiency in energy conversion, high force density, better lifetime, and high stability for CP actuators to use in bioelectronics and biomimetic devices.

The actuation mechanism of the CP actuator is similar to that of natural muscles. In both cases, the actuation resulted from the ion exchange process during the electrochemical reaction triggered by an electric pulse. The insertion/expulsion of ions causes a volume change in both cases used to perform the mechanical work. Many CPs have been shown to be biocompatible and very low electric potential is enough to cause redox electrochemical process/actuation which are the important factors if biomimetic devices are to be incorporated into a living system. The synthesis of CPs are facile. Electrodeposition enables to control of properties and shapes for achieving micro and macro actuators. Numerous facile pattering of CPs are reported to achieve device with specific applications. CP actuators induce relatively large force. The high force density is essential for practical application. The reported force density values for CP actuators are much higher than those reported for natural muscles (<1 MPa). The strain or displacement is another important parameter of an actuator. CP actuators produce relatively large strain. The challenge is to create actuators capable of producing large force along with large displacement.

One of the limitations of CP actuators is slow response time. It depends on the ion diffusion process during the electrochemical reaction. So, the response time can be improved by increasing the porosity of CP film, which improves the ion diffusion, by controlling the synthetic strategy and synthetic conditions. Moreover, the actuation performance is also improved by increasing the ionic conductivity of the electrolyte medium. The degradation of CP limits the lifetime of actuators. The electrochemical overoxidation and the presence of oxygen are responsible for degradation. The overoxidation can be avoided by controlling the applied potential and degradation due to the presence of oxygen can be eliminated by CP actuator encapsulation. Further, the low linear strain of CP becomes a barrier for the construction of an effective linear actuator. The poor mechanical strength such as brittleness, poor elongation at break, and poor processibility constitute major obstacles to its extensive applications [86, 87]. It can be overcomed by incorporating other fillers that offer mechanical strength without having too much sacrifice on conductivity that will affect actuation performance [88–90].

9.3   Conducting Polymer/Hydrogel Systems as Actuators

9.3.1   Importance of Conducting Polymer/Hydrogel Systems

The effective combinations of CPs and hydrogels towards the fabrication of actuators seem to be noteworthy due to the fact that both materials share a common feature of stimuli-responsive behavior, i.e., they can change their shape, size, volume, etc., in response to external stimuli. However, the actuators based on CP and hydrogel combinations are not much explored. With the advent of fabrication techniques and composite formulations, researchers have succeeded to overcome many of the limitations associated with both CPs and hydrogels. The CP/hydrogel systems drawn much attraction to the scientists and researchers as it possess the valued properties of both CPs and hydrophilic polymers such as conductivity, electroactivity, mechanical strength, swelling property, biocompatibility, porosity, high specific surface area, controlled morphology, electrochemical switching between redox forms of CP, and macroscopic homogeneity [7, 91–94]. The first CP/hydrogel system reported was by Gilmore *et al.* in 1994 [95]. It is a hybrid composite of PPy and polyacrylamide in which PPy is directly electropolymerized on a preformed hydrogel.

9.3.2   Conducting Polymer/Hydrogel-Based Actuators in Literature

Table 9.1 shows the various CP/hydrogel systems reported in the field of actuators.

9.3.3   Fabrication of Conducting Polymer Hydrogels

Several successful routes have been proposed for the fabrication of CP/hydrogel systems. The properties of these CP/hydrogels are obviously dependent on the concentration and structure of the CP within the hydrogel. It is essential to say that the preparation procedure and synthetic conditions have a strong effect on the properties of the material that has to be synthesized. It becomes a hurdle in the reliable and

Table 9.1 CP/hydrogel actuators in literature.

System	Type of actuation	Morphology of actuator	Characteristics	Ref.
Chitosan/PANi	Electrochemical bending actuation/self-oscillatory actuator.	Film/membrane	pH < 7: bend towards the cathode, pH > 7: bend towards the anode. At neutral electrolyte-oscillatory bending	[96]
Chitosan/PANi/CNT	Dual mode actuation (chemical and electrochemical actuation)	Microfiber	pH actuation strains: 2%–2.5%, electrochemical actuation strains: 0.3%	[97]
Chitosan/PANi	Chemical and Electrochemical linear actuation	Microfiber	Conductivity: 2.856×10^{-2} S/cm Maximum actuation strain: pH actuation: 6.73%, electrochemical actuation: 0.39%	[98]
PVA/PANi	Electrochemical actuation	Microfibrous mat and rolled up mat	Conductivity: 2.35 S/cm. Maximum actuation strain: 1.8%	[99]
PAAc-co-PVSA/PANi	Electrochemical actuation	Film	Change in volume when subject to an electric field	[100]

(Continued)

Table 9.1 CP/hydrogel actuators in literature. (*Continued*)

System	Type of actuation	Morphology of actuator	Characteristics	Ref.
PANI/PVA	Chemical actuation	Film	Maximum bending angle- 180°–250° and response time 12–35 s (depends on PANi wt.%)	[101]
Cellulose-PANi	Electro actuator	Thick film	Linear crawling motion, maximum crawling velocity : 15.1 mm s^{-1}, ionic conductivity: up to 7×10^{-2} S cm^{-1}, compressive strength: up to 0.48 MPa	[102]
Cellulose acetate/PANi	Electro actuator	Electrospun membrane	Electrically driven bending Maximum ionic conductivity: 9.23×10^{-4} S/cm	[103]
Poly (NIPAM-co-AAc)-PANi	Electromechanical actuation	Thick film	Application of an electric field produces contraction stress than that generated by thermosensitivity	[104]
PVA- PANi	Shape memory performance	Strip	Thermal-, water-, and NIR light-induced shape-memory performance Tensile strength ~83 MPa High recovery stress ~ 6.0 MPa	[105]

(*Continued*)

Table 9.1 CP/hydrogel actuators in literature. (*Continued*)

System	Type of actuation	Morphology of actuator	Characteristics	Ref.
Chitosan-g-PANi	Reverse actuator (Mechano electric actuator)	Film	Mechanic stimulus gives an electric response, Conductivity- in the order of 10^{-2} $S \cdot cm^{-1}$	[106]
PPy/acrylamide/carbon black	Electrochemical bending actuation	Cylinder	Maximum bending angle- $34 \pm 1°$	[107]
PPy/CNT/DNA Hydrogel	Electrochemical linear actuation	Fiber	Actuation stability ~4.41% Electrochemical capacitance ~371 F/g	[108]
PPy/chitosan	Electrochemical bending actuation	Microfiber	Maximum strain: 0.54%. Conductivity: 3.1×10^{-1} Scm^{-1}	[109]
PPy/PEG/alginate hydrogel	Electrochemical actuation	Film	Maximum stress: 1001.0 MPa	[110]
PPy/poly(2acrylamide-2-methyl propane) sulfonic acid (PAMPS) gel	Electrochemical bending actuation	Cylinder/rod	Conductivity: 2×10^{-4} Scm^{-1} Mechanical strength: 300 KPa	[111]

(*Continued*)

Table 9.1 CP/hydrogel actuators in literature. (*Continued*)

System	Type of actuation	Morphology of actuator	Characteristics	Ref.
PPy/PVA-g-P(AMPS-co-AN) gel	Electrical bending actuation	Film	Conductivity: 21.1×10^{-2} Scm^{-1} with a high concentration of PPy	[112]
PTh/CS/CMCS	Bending electroactuation	Film	Maximum bending angle: 70–80°	[113]
PTh/PVA	Electroactuation	Film	Maximum bending angle: 10.61°	[114]
PEDOT/PVDF-*graft*-PEG/ PEGMA	Bending electroactuation	Membrane	Air actuation	[115]
PEDOT/agarose	Electrochemical actuation	Film	Peel off from microelectrode by actuation	[116]
PAAm/PEDOT-PSSs double network hydrogel	Electrochemical bending actuation	Film	Fracture stress: 1.2 MPa, Fracture strain: 90%	[117]

repeatable fabrication of materials [118–122]. Some of the important fabrication methods are given below.

9.3.3.1   Polymerization CP Within a Prefabricated Hydrogel Matrix

The most common method of preparation of a CP/hydrogel is *in situ* polymerization of CPs through preformed hydrogel matrices. The fabricated dried hydrogel is allowed to swell in a CP monomer solution and then the CP is polymerized *via* exposure to a chemical oxidant (chemical polymerization) [123–125]. The reverse strategy, the oxidant imbibed hydrogel exposed to monomer solution, also frequently reported [126–128].

The electrochemical polymerization technique can also be performed to obtain composites instead of using a chemical oxidant, by using a hydrogel coated electrode as anode. The coated electrode was immersed for a few minutes in an electrochemical polymerization bath prior to the electrodeposition of CP by the application of electric potential.

A novel chemical route, called interfacial polymerization, has also adopted for the synthesis of CP hydrogels. This process is carried out in the two-phase system, i.e., monomer and the oxidant ion are in two immiscible liquid phases. Here, the hydrogel is allowed to swell in an aqueous oxidant solution until attaining equilibrium and then exposed to monomer solution in an organic solvent like hexane, and the polymerization of CP takes place at the interface of two liquids which directly diffused into a hydrogel matrix [129, 130].

9.3.3.2   CP-Hydrogels Formed by Blending of CP with Hydrogel

It is an alternative approach to prepare CPHs, which comprises of blending CP with another polymer in solution or in the melt state [131]. It involves mixing the solution of CP with the solution of hydrogel with constant stirring to disperse CP uniformly throughout the matrix polymer along with crosslinker if needed [96, 102, 131–133]. Since CPs shows poor solubility in common solvents, it limits its versatility.

9.3.3.3   Polymerization of CP in the Presence of Matrix Polymer Solution/Dispersion

It is a special type of one-pot synthesis that constitutes the preparation of CP in the solution of a suitable supporting polymer instead of polymerizing through the prefabricated corresponding hydrogel. It is well known that the polymerizations of aniline/pyrrole in the presence of water-soluble polymers, often in presence of steric stabilizers, yield colloidal dispersions of the composite. It can be precipitated with a suitable solvent like methanol, ethanol, ethanol/ether mixture, and acetone or can be cast into a film or fibers [134, 135]. Alternately, the freezing-thawing method combined with *in situ* polymerization of aniline in the acidic aqueous solution of PVA was reported to enhancing mechanical strength [136].

9.3.3.4 CP-Hydrogels Formed from Mixed Precursors

This is an alternative, less commonly used method for the fabrication of CP/hydrogel systems. In this method, the CP and hydrogel precursors are placed in the same vessel where they are polymerized either simultaneously or in a two-step process. This method is restricted to those hydrogels which can be easily prepared from their monomer [137].

9.3.3.5 Synthesis of CP Gels by Cross-Linking with Dopant Molecule

Conventional methods of fabrication of conductive polymer hydrogels usually result in a gel system that consists of conductive component CP within the nonconductive components, thus deteriorating the conductivity of the resulting material. Apart from these methods, it is found that molecules with multiple acid functional groups, such as phytic acid, copper phthalocyanine-3,4',4",4'''-tetrasulfonic acid tetrasodium salt (CuPcTs), and amino trimethylene phosphonic acid (ATMP) can cross-link the conductive polymer chains, leading to CP gels with 3D networked structures free of insulating components with good electronic and electrochemical properties [138–151]. In order to produce CP gels, the two-component mixing strategy was adopted: a solution containing the oxidative initiator was mixed with a solution containing the CP monomer and crosslinker dopant. The resulting solution gelled to form a hydrogel within a few minutes [138, 140].

9.3.4 Polyaniline/Hydrogel Systems as Actuators

A self-oscillatory actuation at constant DC voltage with pH-sensitive chitosan/polyaniline hydrogel was demonstrated by Kim *et al.* [96]. The chitosan/polyaniline membrane shows electric potential induced bending actuation. The degree of bending increases with increasing the magnitude of the applied DC voltage and time, and it reaches the maximum value in a few seconds when measured in the pH 1 buffer solution (Figure 9.4a).

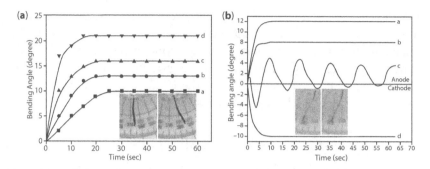

Figure 9.4 (a) Bending degree of film for various applied potentials [(a) 3, (b) 5, (c) 7, and (d) 10 V] at pH 1 and photographs show initial position (left) and maximum bending at 5 V (right). (b) Bending degree at different pH (applied potential = 5 V); (a) pH 1, (b) pH 4, (c) pH 7, and (d) pH. Picture show limits of small oscillatory motion at pH 7. Reprinted with the permission from [96], copyright of American Chemical Society.

The direction and magnitude of bending also vary as the pH of the medium changes. When CP membranes subjected to an applied electric field in acidic solution (< pH 7), they bent toward the anode, while in a basic solution (> pH 7) bent toward the cathode. However, an oscillatory bending was observed in a neutral electrolyte (Figure 9.4b).

The pH depended on the bending mechanism of the chitosan/ polyaniline membrane shown in Figure 9.5. In alkaline electrolytes where the polyaniline is not electroactive, but a local pH change due to water

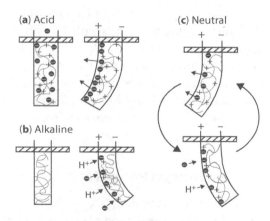

Figure 9.5 The mechanism for bending actuation of chitosan/polyaniline membranes in different pH electrolyte solutions. (a) In acidic electrolytes, bending towards the anode. (b) In basic electrolyte, bending towards the cathode. (c) In neutral electrolyte, a cyclic oscillation in bending. Reprinted with the permission from [96], copyright of American Chemical Society.

hydrolysis generates local acidification at the anode side, causing protonation of the chitosan and polyaniline, swelling at the anode side of the membrane resulted in a net bending toward the cathode. In acid electrolytes, the polyaniline electrochemistry contributes to the bending toward the anode. In acid electrolyte, the initial high charge density (+) in membrane induces the motion of mobile anions to the anode causes a charge shielding and shrinkage. Furthermore, the oxidation of PANi at the anode leads to anion expulsion, causing additional shrinkage at the anode. In neutral electrolyte, the initial movement is toward the anode because of charge shielding effects (as in acidic solution) and subsequent protonation of the polymer at the anode causes local swelling leads to bending back toward the cathode (as in basic solution). The protonation renders the PANi becomes electroactive by protonation, so that oxidation may occur at the anode, causing shrinkage that promotes bending back toward the anode. Therefore, a cyclic oscillation in bending results.

A detailed study in a volume change of polyacrylic acid-*co*-poly(vinyl sulfonic acid)/polyaniline interpenetrating polymer network (IPN) hydrogels conducted by Kim *et al.* [100]. When IPN hydrogels subjected to an electrical stimulus, hydrogel underwent a very sensitive change in volume. A reversible change in volume is developed on switching the electrical stimulus between the positive and negative electrodes by contraction and expansion of the IPN hydrogels, respectively. The magnitude of the change in volume of this hydrogel is proportional to the applied voltage. So, electrical stimulus-responsive behavior is important in actuator applications.

A "dual mode" actuation was reported for the first time in chitosan/polyaniline/carbon nanotube fibers fabricated using a wet-spinning method by Spinks and coworkers in 2006 [97]. It includes pH actuation which produces a large strain due to the protonation/deprotonation of amine groups of chitosan present in the fiber and electrochemical actuation strain response due to the redox reactions of the PANi.

Strain profile of chitosan/PANi/SWNT fiber during pH switching between pH 1 and 7 and pH 1 and 4 shown in Figures 9.6a and b, respectively. The fiber expands at pH 1 and contracts at pH 4 and 7 in both cases. The expansion of chitosan at low pH attributed to the ionic repulsion of the protonated amine groups ($-NH3^+$). While the protonation level decreases in high pH results in contraction [152]. The system attains the equilibrium state more readily at pH 7 than at pH 4. The measured strains were 2.5% and 2%, respectively, which is much greater than reported magnitude of the chemically induced actuation of the PANi, typically 1%–2%. Fiber shows electrochemical actuation in a 0.1M HCl solution within the applied electrical potential –0.2 to +1.0 V (Figure 9.6c). The strain profile shows

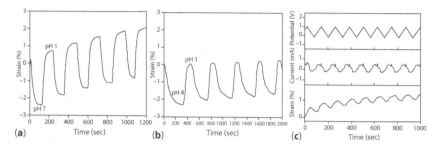

Figure 9.6 Strain profile for the contraction and expansion of chitosan/PANi/SWNT fibers during pH switching between (a) pH 1 and 7 and (b) pH 1 and 4. (c) electrochemical actuation profile of the chitosan/PANi/SWNT. The applied potential, current, and strain in a 0.1M HCl aqueous solution at scan rate 10 mV/s. Reprinted with the permission from [97], copyright of Elsevier.

expansion and contraction during oxidation and reduction respectively. During oxidation of PANi anion inserted into polymeric chain causes expansion and during reduction expulsion of anion resulted in contraction. The electrochemical strain measured is 0.3% much lower than reported for neat polyaniline fibers (1.7%) having much higher conductivities [153], presumably due to the diluting effect of the chitosan and stiffening effect of the nanotubes incorporated in this fiber.

Chitosan/PANi/SWNT fiber produces a "dual mode" actuation by simultaneous pH switching and electrical potential cycling (Figure 9.7). The macro and micro strains are attributed to the pH behavior of chitosan and the redox property of PANi present in the fiber, respectively, as explained above. However, at higher pH (i.e., pH 7), a gradual decrease of microstrain is observed (Figure 9.7b) since the PANi gradually loses electroactivity at pH 7. However, at pH 4, the fiber shows consistent microstrain because the PANi in the fibers retains electroactivity at this pH but smaller than that at pH 1 because the protonation level at pH 4 is lower than that at pH 1.

Electrochemical actuation and pH-induced actuation also demonstrated in chitosan/PANI microfibers [98]. The microfibers fabricated by *in situ* chemical polymerization of aniline on chitosan microfibers obtained by the wet spinning method.

Figure 9.8a shows the chemical actuation of this fiber corresponding to a pH switching between pH = 0 and pH = 1 in the HCl solution. It shows a high strain ratio of 6.73%, which is typical for a biopolymer hydrogel under stress, suggesting that the biocompatible nature of the microfibers was maintained even after incorporation of the polyaniline in the chitosan hydrogel matrix. The microfiber elongated at high pH while contracted at low pH under a load of 5.5 mN (stretched fiber), as similar to chitosan fiber

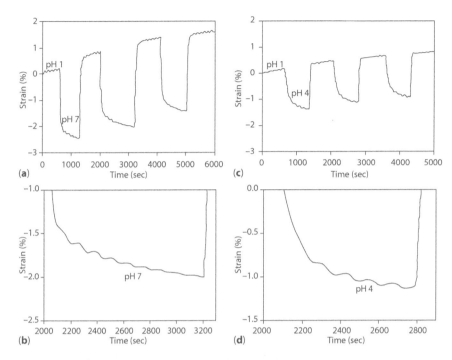

Figure 9.7 Strain profile for dual mode actuation of chitosan/PANi/SWNT fibers during pH switching between (a) pH 1 and 7 and (c) pH 1 and 4, and simultaneous potential window from −0.2 to +1.0V. Magnification of electrochemical actuation in (b) pH 7 and (d) pH 4 (Scan rate: 10 mV/s). Reprinted with the permission from [97], copyright of Elsevier.

under similar conditions. This is opposite to that of a non-stretched chitosan fiber, which generally expands at low pH and contracts at high pH, as in chitosan/PANi/CNT microfibers discussed earlier. The electrochemical actuation profile of the fibers generated by potential cycling between −0.2 and 0.8V in aqueous HCl electrolytes at different pH is shown in Figure 9.8b. It shows a high strain of 0.39% at pH = 0 and strains gradually decreases as the pH increases. The electrochemical actuation due to the redox process of PANi associated with the fiber in cycling potential.

A flexible CP actuator was fabricated using polyaniline coated on electrospun poly(vinyl alcohol) (PVA) nanofibrous mat by Ismail *et al.* [99]. The dual-mode muscle lever arm system with an attached cyclic voltammogram was used to study the electrochemical actuation performance. A single strip of PVA/PANi hybrid mat shows a maximum linear actuation strain of 1.8% during cycling the potential between specified ranges in 1 M methane sulfonic acid solution (Figures 9.9a and b). This hybrid mat could be rolled-up into a multilayered assembly of cylindrical structures (Figure 9.9c). These

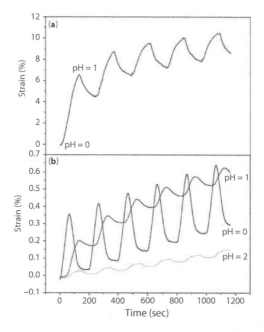

Figure 9.8 Strain profiles of chitosan/PANi microfibers in HCl during: (a) chemical actuation during pH switching between pH = 0 and 1, and (b) electrochemical actuation at different pH (Scan rate: 10 mV/s). Reprinted with the permission from [98], copyright of Elsevier.

rolled-up structures provide a huge surface area per unit volume due to interlayer spacing. It contributes towards easy diffusion of the electrolyte for an efficient electrochemical reaction and leads to enhanced actuation performance and fast response because current produced during the electrochemical reaction is dependent on the active surface area. The rolled-up system produces a linear actuation strain of 3.67% (average strain = 3.2%) under almost freeload. It can able to tune the actuation performance by controlling the number of times it was rolled. Figure 9.9d shows the variation in actuation strain of rolled-up structure with scan rate that follows an Arrhenius exponential behavior, and corresponding response speed directly proportional to the scan rate. It points out that an actuation strain of 0.2% was achieved in few seconds (5 s) in the rolled-up structure. Therefore, suggesting that the designing of an actuator based on CP having a moderately high strain with a fast response is possible.

Zhang *et al.* reported a monolithic rapid reversible actuator responding to acid/base fabricated using polyaniline (PANI) nanofibers and polyvinyl alcohol (PVA) nanocomposite [101]. The flash-welding technique is adopted to form PANI/PVA films with asymmetric structure. The flash

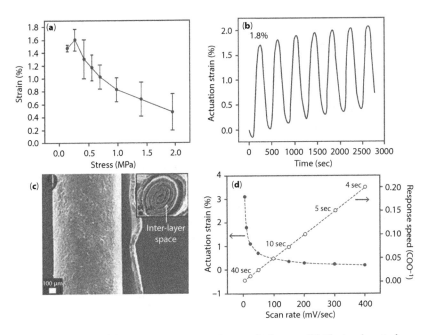

Figure 9.9 (a) Plot of the actuation strain against applied stress. (b) Electrochemical actuation profiles at 0.28 MPa. (c) FE-SEM image of the rolled-up structure of the PVA/PANI hybrid mat. (d) The plot showing the variation in strain and the subsequent response speed with increasing scan rate of roll-up structure. Reprinted with the permission from [99], copyright of Elsevier.

welding at the surface produces a smooth, continuous surface that differs from the porous bottom part (Figure 9.10). The flash welding decreases doping sites of PANi that is caused by cross-linking or degradation, induced by flash heating. Therefore, upon protonic acid doping, large swelling occurs to the non-welded sides of films than welded sides. This asymmetric volume change resulted in the bending of the film towards the welded side in acid medium. It regains to the original position in basic medium.

The PANI/PVA nanocomposite actuator with 68 wt% PANI bended more than 180° within 18 s in 1 mol/L camphor sulphonic acid (CSA) solution, and after rinsed in water for 14 s the film recovered to 90°, and it took only 1.2 s to recover to its original flat state in 1 mol/L NaOH solution. Figure 9.11a shows bending angle-time curves of PANI/PVA nanocomposite actuators with different PANI contents. It indicates that the controllable bending angles can be achieved by tuning PVA contents in films. Moreover, the actuation rate is also highly pH-dependent (Figure 9.11b). The actuation rate sharply decreases with the increase of pH of the acid medium. It suggests that the controllable responsive rate can be achieved

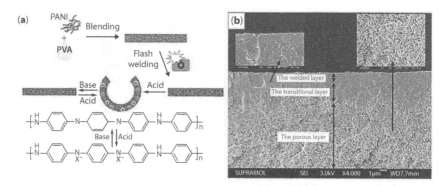

Figure 9.10 (a) Schematic diagram of the preparation process and the actuation mechanism. (b) A cross-sectional SEM image of a flash-welded PANI/PVA film shows about a 1 μm welding layer on a glass substrate. Reprinted with the permission from [101], copyright of Elsevier.

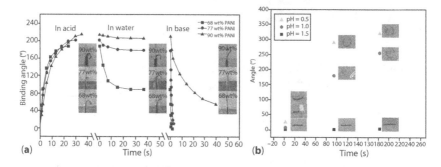

Figure 9.11 (a) Curve shows time-dependent bending angles for PANI/PVA actuators with different PANI contents actuated by 1 mol/L CSA solution, water, and 1 mol/L sodium hydroxide solution, respectively; images of corresponding ultimate bending states of each stage are given. (b) Actuation of PANI/PVA film in three CSA solutions (pH 0.5, 1.0, 1.5) figure showing different degrees of bending at the same time. Reprinted with the permission from [101], copyright of Elsevier.

by changing the pH of the external media. Besides, a monolithic PANI/PVA boxes with tunable folding/unfolding movement in acid/base also demonstrated *via* proper modulation using asymmetric strips as hinges between faces.

Shi *et al.* demonstrated a linear crawling motion under the applied electric field of 26.6 V using PANI–cellulose composite hydrogels prepared with 25 wt% PANI content in the dried hydrogel [102]. The composite hydrogels moved from negative to positive as shown in Figure 9.12. The possible response mechanism for this behavior could be explained as follows. When the doped PANI–cellulose hydrogels placed in an electric

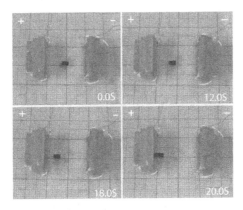

Figure 9.12 Time profiles of PANI–cellulose composite hydrogels in motion under the applied electric field. Reprinted with the permission from [102], copyright of Royal Society of Chemistry.

field, a local high ionic strength was developed near the end of the positive electrode due to the migration of mobile anions and the positive charges remain on the polymer backbones. This electrostatic shielding resulted in swelling or shrinking of the material. Besides, hydrogels experiencing an electrostatic force along the applied electric fields due to the difference in the high ionic strength at the two ends of the material. When this electric field exceeds a certain critical driving threshold, crawling motion of the doped PANI–cellulose hydrogels could be observed from the negative to the positive electrode. Crawling velocity increases as the driving time increases because of the electrostatic force increases as a decrease in the distance between the hydrogels and the positive electrode. The maximum crawling velocity found up to 15.1 mm s^{-1}.

Hong *et al.* developed a dry type electro-active biopolymer actuator with electrospun cellulose acetate-polyaniline (PANi/CA) bio-composite membranes [103]. The harmonic responses for this composite membrane responding to sinusoidal electrical inputs with an excitation frequency of 0.1 Hz and the voltage amplitude of 3.0 V used to establish the actuation behavior. The tip displacement of the PANi/CA bio-composite actuator is much larger than that of the pure CA actuator. Even the minute level of PANi in cellulose acetate increases the bending actuation performance of cellulose acetate nanoporous bio-composite actuators.

da Silva prepared a semi-interpenetrating network of poly(isopropyl acrylamide-co-acrylic acid)/polyaniline hydrogel and tested the capacity of the hydrogel to generate stress, through controlled changes in the temperature and using an electrical stimulus [104]. This temperature-sensitive hydrogel can able to generate contraction stress when a temperature change applied over a wide range of values. The application of an electric field to these systems generates contraction stress other than that generated by thermosensitivity, mostly due to the generation of heat through the Joule effect.

A multi-responsive shape memory composite prepared by dispersing polyaniline fibers into polyvinyl alcohol (PVA) by *in situ* polymerization reported very recently [105]. The PANi fibers increase additional physically cross-linking points in composites through hydrogen bond and also serve as photothermal conversion reagents, which results in excellent water-, thermal-, and near-infrared (NIR) light-induced shape memory properties. NIR-induced shape recovery ratio and speed could be enhanced *via* the increase of PANi percentage in composite and the light power density. Besides, the composites show high recovery stress over 6.0 MPa and it increased with the increase of temperature and polyaniline loading percentage.

The temporary "m" shape samples were obtained through the same process as above. In Figure 9.13a, the sample was placed on a thermal platform

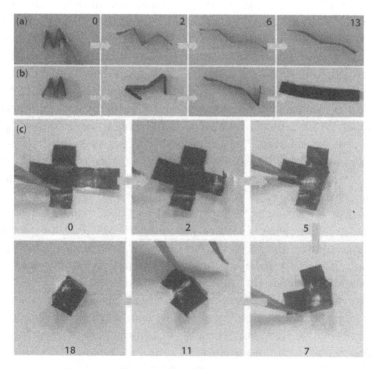

Figure 9.13 (a) Thermal-induced shape recovery of PVA-PANi composite (composite contains 10 wt% PANi with respect to PVA) at 80°C; (b) and (c) NIR light-induced shape memory behavior under 2.9 W cm−2 of NIR light irradiation. Reprinted with the permission from [105], copyright of American Chemical Society.

at 80°C and its original shape could be recovered within 13 s, indicating its good thermal-induced shape-memory performance. In contrast, the NIR light-induced shape recovery process in Figure 9.13b showed that the original straight shape was recovered in three steps by locally irradiating the sample. In addition, a 3D shape structure (cube) was used to demonstrate the local control of the NIR light-shape recovery process (Figure 9.13c).

In contrast to the above discussion, Marcasuzaa *et al.* studied the ability of the chitosan-*graft*-polyaniline-based hydrogels to convert mechanical work into electrical energy as a "reverse actuators" [106]. This composite hydrogel produces an electric response to a mechanic stimulus, i.e., pressure.

9.3.5   Polypyrrole/Hydrogel Systems as Actuators

Lewis *et al.* have studied the bending actuation of polyacrylonitrile/PPy and polyacrylamide/PPy composites [154]. The hydrogel polyacrylonitrile (PAN) and polyacrylamide (PAAm) were developed as a solid polymer electrolyte (SPE) and PPy/PAN/PPy and PPy/PAAm sandwich structures were fabricated for actuator applications. Due to slow ion movement and stiffness, SPE decreases the rate of actuation when compared with that in solution. But it offers the possibility of such actuators as free-standing devices in the air. In a study by Yamauchi *et al.*, a cylindrical bending actuator from composite materials of PPy and poly (2-acrylamide-2-methylpropane) sulfonic acid (PAMPS) gel was developed through chemical oxidative polymerization. This gel bent towards the cathode by the application of the electric field and the bending rate depends on the ratio of PPy in the composite [111].

Moschou *et al.* have developed a cylindrical bending electroactuator that actuated at near-neutral pH conditions by the application of low potential (3 V or less) in a reversible manner. The artificial muscles are composites of acrylic acid and acrylamide doped with conductive additive polypyrrole/carbon black [107]. The cylindrical sample is placed in an electrochemical cell between two Pt electrodes and the electroactuation is expressed in terms of the hydrogel bending angle and it possesses a parabolic shape bending towards the cathode (Figure 9.14). The actuation is explained by the theory of Flory [155] and Tanaka [156]. According to this, the changes in the osmotic pressure due to ionic distribution will result in electroactuation. The ion concentration gradient between the inner and outer parts of the hydrogel results in electrophoretic movement and causing the bending actuation. Thus, higher bending can be achieved by increasing the ionic distribution within the hydrogel.

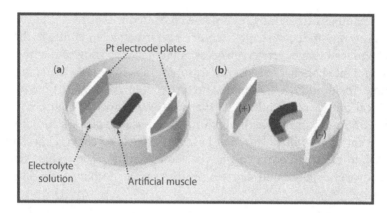

Figure 9.14 Artificial muscle sample placed between two Pt electrodes in a 0.15M NaCl solution by the application of 3V for a time period of 2 min (a) no electroactuation and (b) electroactuation. Reprinted with the permission from [107], copyright of Elsevier.

Lee *et al.* have reported a simultaneous electrochemical linear actuation of polypyrrole/carbon nanotube hybrid fibers doped with porous structured DNA hydrogel [108]. The fibers are prepared by simple chemical polymerization. As there occurs hydrogen bonding interaction between DNA hydrogel and polypyrrole, it enhances the electrochemical and mechanical stability of the CP [157, 158]. DNA is highly porous, negatively charged, and acts as counter anion for charge balancing during the polymerization of PPy. During reduction, the negative charge of anionic DNA/CNT is balanced by the incorporation of small cations and during oxidation, these cations are expelled [159]. In PPy/CNT layer, the actuation is restricted by the inactivated inner core due to insufficient number of pores (Figure 9.15) [160]. DNA possesses extended pore structure, and hence, the ions can transport throughout the internal volume. Therefore, the DNA/Ppy/CNT hybrid fibers show improved actuation stability with expansion and contraction of ~4.41% in low concentrated aqueous electrolyte solutions at low potential.

Ismail *et al.* have reported polypyrrole/chitosan microfiber for artificial muscle applications. The microfiber was fabricated through the wet spinning of the chitosan solution followed by the *in situ* chemical polymerization with pyrrole [109]. It also acts as a sensor of working conditions such as applied current, temperature, and electrolyte concentration with an electrical conductivity of 3.1×10^{-1} Scm^{-1}. The sensing ability and electroactivity are provided by polypyrrole, and chitosan is responsible for the mechanical strength and swelling/shrinking processes. The electrochemical

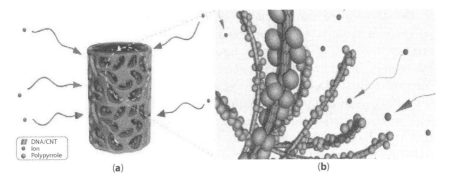

Figure 9.15 Schematic representation of ion diffusion in a DNA/PPy/CNT hybrid system during actuation (a) view of the outer surface and (b) ion diffusion on the inner surface of the DNA/PPy/CNT hybrid fibers. Reprinted with the permission from [108], copyright of Elsevier.

actuation of the fiber was studied using a dual mode muscle lever arm and a linear actuation strain of 0.54% was achieved upon cycling the potential in an aqueous system. A schematic representation of the linear actuation of CS/PPy microfiber is presented in Figure 9.16. The central part provides less actuation than the external part as the concentration of PPy is higher at the surface of the microfiber. The sensing characteristics were monitored using chronopotentiogram. The consumed energy during the electrochemical reaction was a linear function of the applied current and working temperature suggests it as a current sensor and temperature sensor, the logarithmic dependence of the electrolyte concentration suggests it as a concentration sensor.

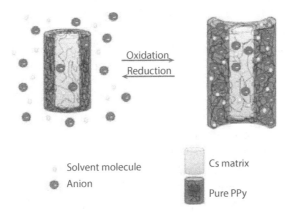

Figure 9.16 A schematic representation of the linear actuation of Cs/PPy microfiber. Reprinted with the permission from [109], copyright of Elsevier.

The electrical bending actuation of a composite material consisting of polypyrrole nanoparticles impregnated into poly(vinyl alcohol)-g-poly(2-acrylamide-2-methyl-1-propane sulfonic acid-co-acrylonitrile) hydrogel has been reported by Mahloniya *et al.* [112]. Wang *et al.* have synthesized composite films of alginate hydrogel and polypyrrole/polyethylene glycol (PPy/PEG). This composite is electroactive and characterized as actuators with improved stress-strain performance as compared to the PPy/PEG film [110]. The biocompatible alginate hydrogel is responsible for the mechanical reinforcement of the film.

9.3.6   Polythiophene/Hydrogel Systems as Actuators

As PANi and PPy, Polythiophene (PTh) also have the ability to behave as an actuator by external stimuli. It is a good conducting, biocompatible electrode material for the fabrication of actuators. Pattavarkorn *et al.* reported the first polythiophene actuator with chitosan and carboxymethane chitosan. After that, scientists put their interest in thiophene derivatives like 3,4-ethylene dioxythiophene. Recently, Poly(3,4-ethylenedioxy thiophene) (PEDOT)–based actuators got more attention.

The freestanding electrode of thiophene/hydrogel is very rare due to its difficulty in fabrication. Pattavarkorn *et al.* reported, in 2013, polythiophene chitosan/carboxymethyl chitosan (PTh/CS/CMCS) composite freestanding film in the motive of developing actuator [113]. They synthesized polythiophene separately by the chemical oxidation method in chloroform using $FeCl_3$ as an oxidizing agent [161]. Chitosan and chitosan derivative (carboxymethyl chitosan) used as hydrogel [162]. Electroactive performances of the fabricated hydrogel were checked with bending actuation by an external electric field. DC electric field is applied and the bending angle is calculated from the equation $\theta = \arctan (A/B)$.

Bending actuation of CS/CMCS, PTh/CS/CMCS hydrogel is the evidence of electroactive behavior of them. The contraction and expansion were happened by the polarization of the amino group of chitosan by the application of the external field. The hydrogel electrode bent towards the cathodic side by the external voltage, it is because of the NH_3^+ group move towards the anodic side. They have studied the changes in bending angle with respect to volume ratio between CS: CMCS (Figure 9.23), the concentration of cross-linking agent (Figure 9.22), the strength of the electric field (Figure 9.21), and presence of polythiophene in hydrogel system.

Pattavarakorn *et al.* also reported that polythiophene/PVA composites films for actuation [114]. Polythiophene/PVA composite is prepared by mixing both thiophene monomer and PVA: CS in the ratio (75:25) using glutaraldehyde as the cross-linking agent. The resulting free-standing film can actuate by an external voltage. The bending response and electrorheological studies of the materials were studied in an external electric field. The change in the bending angle and the volume variation with respect to the particle concentration and strength of the electric field also were investigated. A silicon oil bath was used for the analytical setup under the DC field with a strength of 0–700 V/mm. The bending angle varied with the concentration of polythiophene concentration in the hydrogel composite. There is an increase in the bending angle with an increase in the concentration of the polythiophene particle. The bending angle with respect to the field-effect also determined. The angle increases with the increase in strength of the field.

PEDOT: PSS/hydrogel system is also reported. Soichiro Sekine *et al.* reported the PEDOT/agarose system for actuation [116]. Here, they fabricate the material through micropatterning of PEDOT on the hydrogel agarose, by two electrochemical processes [164]. The first one is the electropolymerization of PEDOT into the agarose skeleton and the other is electrochemical actuation assisted peeling (Figure 9.17).

They introduced a printing method on the moist gel. From Figure 9.17, they used platinum microelectrode and pour the melted agarose on it. Electropolymerization of PEDOT has been done on this agarose. For peeling, electrochemical actuation had done several times. The volume of

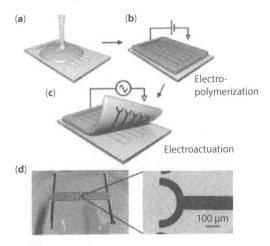

Figure 9.17 (a–c) Schematic diagram of conducting polymer/hydrogel system fabrication. (a) The agarose (melted)poured on platinum microelectrode. (b) Electropolymerization of PEDOT on agarose hydrogel/Pt electrode. (c) Peeling of coated film from Pt electrode by electrochemical actuation. (d) Photograph of PEDOT/agarose microelectrode. Reprinted with the permission from [116], copyright of American Chemical Society.

PEDOT deposited on the hydrogel template is increased by consecutive redox cycle. The resistance of the PEDOT printed surface can be analyzed by the four-probe method in wet conditions. It can store in the water for long days. Eletrochemical actuation (0.5 V vs. Ag/Ag/AgCl) is an effective method to peel the film from the master electrode than vigorous peeling. Otherwise, the film/soft gel will collapse. *In vivo* lapping electrode and *in vitro* cell cultivation are the unique applications of this flexible electrode. They can mimic biological properties. Here, Soichiro Sekine *et al.* demonstrated the moving flexible electrode by electric stimulus support for the myotubes. This condition is similar to the glucose uptaken by the skeletal muscle. Electric stimuli motivate the actuation of the PEDOT/agarose electrode and this will induce the contraction of the muscle cell. This technique can be applied for other hydrogels such as collagen and fibrin, etc.

Simaite *et al.* have been reported that the hydrogel graft membrane improves the strength of PEDOT: PSS/PVDF/Ionic liquid actuators. PVDF surface has been functionalized by the hydrogel PEG or PEGMA (Polyethylene glycol methacrylate) [115]. PVDF surface has been modified by graft polymerization of PEG or PEGMA. This leads to the formation of the PVDF-*graft*-PEG/PEGMA. Tri-layer actuator was formed by the drop-casting of PEDOT: PSS system on the fabricated graft polymer. The formed rectangular actuator can actuate bidirectionally by applying a voltage from one side. The actuation is performed in the presence of ionic liquid. PEDOT: PSS/PVDF system is hydrophobic, the membrane became hydrophilic after functionalized with PEG. PVDF-*graft*-PEG/PEGMA membrane with PEDOT: PSS system can perform air actuation for more than 150 hours and more than 50,000 cycles.

PEDOT: PSS hydrogel system for actuation had reported by Dai *et al.* They have used triple-network and special double network hydrogels contain poly(acrylamide) (PAAm) and poly(acrylic acid) [117]. The hydrogels are used here for strengthening the system. Here, they introduced physically cross-linked CP network. It was obtained by the supramolecular self-assembly. They focused on PEDOT: PSS/PAA/PAA triple network and PEDOT: PSS/PAAm special double network. Both have good mechanical and electrical properties. The swelling-deswelling properties of the material have investigated under a change in pH, ionic strength, and electric field. The negative charge in the PAA screen by the PEDOT network from the electric field. So, the triple hydrogel network cannot deform by an external electric field. PAAm special double network can distort by the same. The PEDOT: PSS/PAAm hydrogel system enhances the electrochemical actuation. The negatively charged carboxylate group can be formed by the hydrolyzation of amino groups in PAAm. By this method, the hydrogel system can respond to the electric field. It is because of the change in osmotic pressure by the electric field. This results in an increase in conductivity, which improves the electrochemical actuation. In summary, they explained self strengthened PEDOT: PSS/hydrogel systems for actuation and other applications.

Ingnas *et al.* reported water-swollen PEDOT: PSS cross-linked CP gel for actuation [165]. This material can swell in water, formed by the self-assembly of the particle. Electrode material has fabricated by PEDOT: PSS hydrogel associated with polypyrrole. During the oxidation-reduction process, the volume will change. The bending force does not occur with smaller anions. The anions will compensate for the charge by the volume expansion of the electrode material. Larger anions (DBS) created bending force within the material.

9.3.7 Factors Affecting Actuation

Polymerization temperature: The electroactuation of a CP/hydrogel system can be increased by increasing the polymerization temperature. As the polymerization process follows the mechanism of free radical polymerization, an increase in temperature will increase the rate of decomposition of the initiator. As a result polymer chain with shorter lengths is formed [166]. The resistance to bending decreases as the chain length of polymer decreases [167], and hence, electroactuation response increases.

The concentration of conducting polymer: The bending behavior of the hybrid system depends on the impregnation of CP within the hydrogel matrix. The effective bending angle increases with increasing CP concentration within the gel up to a certain percentage impregnation of CP [112]. By the introduction of CP into the hydrogel matrix, there occurs an unsymmetrical distribution of charge and hence induces ionic mobility within the hydrogel. Thus, higher the CP concentration, the higher would be the bending response [113, 114]. However, if the concentration of CP increases beyond a certain percentage, the bending angle decreases due to the hydrophobic nature and rigidity of CP.

Effect of voltage: The bending behavior of the CP/hydrogel system is observed only above a certain voltage called a lower critical voltage. Above the lower critical voltage, as voltage increases, the attraction of the charged matrix towards the electrode increases and hence bending response increases [96, 112].

Electric field strength: As the electric field strength increases, the bending response and hence the bending angle of the actuator increases (Figure 9.18) [113]. As an electric field is increased, the CP gets more

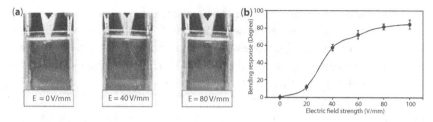

Figure 9.18 Dependence of bending behavior of cross-linked 3:2 CS/CMCS hydrogel on electric field strengths. Reprinted with the permission from [113], copyright of Elsevier.

polarized leading to higher dipole moment and intermolecular interactions [114].

Presence of cross-linking agents: The introduction of suitable cross-linking agents results in the development of mechanically stable hydrogel-based actuators with reproducible bending responses. The cross-linking agents form covalent bonds between the monomers and increase the density of the polymer. The cross-linking agents having the ability to form hydrogen bonding with the monomers increases the mechanical stability, enhances their rigidity and thus restricting the possibility for higher electroactuation [107]. The cross-linking agents which do not have hydrogen bond donor groups result in higher electroactuation due to less cross-linking.

The cross-linking agents like glutaraldehyde have the ability to break hydrogen bonds and increase the number of free amino groups [162], thus increases the bending response. As the concentration of cross-linking agents increases, the bending response also increases up to a certain concentration. Beyond that, bending response decreases due to higher rigidity of the hydrogel (Figure 9.19) [113].

Active surface area: The surface area of the CP/hydrogel systems also plays an important role in enhancing its actuation response [99]. As the surface area increases, the actuation behavior increases.

Content of similar charged groups: The higher content of similar charged groups increases the electrostatic repulsion between them, and hence, the chain relaxation increases. As a result, the electroactuation response increases. This is proved by Moschou *et al.* by the introduction of a dicarboxylic acid instead of a monocarboxylic acid [107]. The dicarboxylic acid contains a higher content of negatively charged groups when compared to monocarboxylic acid. Here, the hydrogel sample containing maleic acid possesses a bending angle of 34 ± 1°, and it is much higher than the bending response of 23 ± 0.6° presented by the hydrogel sample based on acrylic acid.

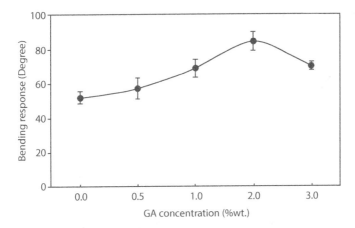

Figure 9.19 Bending response under 100 V/mm electric field of CS/CMCS hydrogel at different GA concentrations. Reprinted with the permission from [113], copyright of Elsevier.

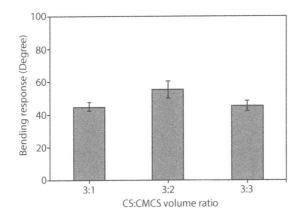

Figure 9.20 Bending angle at various CS: CMCS ratio under 100 V/mm. Reprinted with the permission from [113], copyright of Elsevier.

The ratio of the different hydrogel, if present: If more than one hydrogel is used in the system, then the actuation also depends on their volume ratio. It was reported by Pattavarkorn *et al.* by fabricating polythiophene-chitosan/carboxymethyl chitosan composite [113]. They have studied the change in bending angle with respect to volume ratio between CS: CMCS. The volume ratio of CS and CMCS varies without a cross-linking agent and the bending angle is measured under a 100 V/mm electric field. Here, they reported that the bending angle is limited due to hydrogen bond within the amino group (Figure 9.20).

9.4 Conclusion and Future Outlook

This chapter highlighted the benefits of CP/hydrogel systems over CPs and hydrogels. CP/hydrogel systems are promising materials for actuators. The many positive attributes of CP/hydrogel systems such as electroactivity, conductivity, mechanical strength, biocompatibility, swelling properties, and redox switching have resulted in an ever-increasing scientific interest in this area. The combination of CPs with a host of hydrophilic polymers to form new material systems of synergistic properties highly relevant for actuator applications. This chapter also discussed the various factors that affect CP/hydrogel actuation. Although there are excellent properties for CP/hydrogel systems, they have not yet been fully explored. Due to the complex structures of CP/hydrogel systems, there are still have many challenges in fulfilling their full potential applications. The relationships between the various parameters like ion-type, the concentration of electrolyte, morphology, etc., and the actuation behavior are not clear. Further studies will be carried out to consider these.

References

1. Geryak, R. and Tsukruk, V.V., Reconfigurable and actuating structures from soft materials. *Soft Matter*, 10, 9, 1246–1263, 2014.
2. Ionov, L., Biomimetic hydrogel-based actuating systems. *Adv. Funct. Mater.*, 23, 36, 4555–4570, 2013.
3. Randhawa, J.S. *et al.*, Chemically Controlled Miniature Devices: Microchemomechanical Systems. *Adv. Funct. Mater.*, 21, 13, 2395–2410, 2011.
4. Ionov, L., Polymeric actuators. *Langmuir*, 31, 18, 5015–5024, 2014.
5. Martinez, J.G., Otero, T.F., Jager, E.W., Effect of the electrolyte concentration and substrate on conducting polymer actuators. *Langmuir*, 30, 13, 3894–3904, 2014.
6. Córdova, F. *et al.*, Conducting polymers are simultaneous sensing actuators. SPIE Smart Structures and Materials + Nondestructive Evaluation and Health Monitoring. SPIE, vol. 8687, 2013.
7. Mawad, D., Lauto, A., Wallace, G.G., Conductive polymer hydrogels, in *Polymeric hydrogels as smart biomaterials*. pp. 19–44, Springer, Switzerland, 2016.

8. Naveen, M.H., Gurudatt, N.G., Shim, Y.-B., Applications of conducting polymer composites to electrochemical sensors: A review. *Appl. Mater. Today*, 9, 419–433, 2017.

9. Liu, Y. *et al.*, Physically crosslinked composite hydrogels of PVA with natural macromolecules: Structure, mechanical properties, and endothelial cell compatibility. *J. Biomed. Mater. Res. Part B: Appl. Biomater.: An Official Journal of The Society for Biomaterials, The Japanese Society for Biomaterials, and The Australian Society for Biomaterials and the Korean Society for Biomaterials*, 90, 2, 492–502, 2009.

10. Kumar, A.M. *et al.*, Biocompatible responsive polypyrrole/GO nanocomposite coatings for biomedical applications. *RSC Adv.*, 5, 121, 99866–99874, 2015.

11. Kumar, A.M. *et al.*, Promising bio-composites of polypyrrole and chitosan: Surface protective and *in vitro* biocompatibility performance on 316L SS implants. *Carbohydr. Polym.*, 173, 121–130, 2017.

12. Taşdelen, B., Conducting hydrogels based on semi-interpenetrating networks of polyaniline in poly (acrylamide-co-itaconic acid) matrix: Synthesis and characterization. *Polym. Adv. Technol.*, 28, 12, 1865–1871, 2017.

13. Huang, H. *et al.*, Self-assembly of polypyrrole/chitosan composite hydrogels. *Carbohydr. Polym.*, 95, 1, 72–76, 2013.

14. Reynolds, J.R., Thompson, B.C., Skotheim, T.A., *Handbook of conductiong polymers, 2 volume set*, CRC Press, New York, 2007.

15. Otero, T. and Sansinena, J., Artificial muscles based on conducting polymers. *Bioelectrochem. Bioenerg.*, 38, 2, 411–414, 1995.

16. Entezami, A.A. and Massoumi, B., Artificial muscles, biosensors and drug delivery systems based on conducting polymers: A review. *Iran. Poly. J. (English Edition)*, 15, 13–30, 2006.

17. Jager, E.W., Smela, E., Inganäs, O., Microfabricating conjugated polymer actuators. *Science*, 290, 5496, 1540–1545, 2000.

18. Valero, L. *et al.*, *Characterization of the movement of polypyrrole-dodecylbenzenesulfonate–perchlorate/tape artificial muscles. Faradaic control of reactive artificial molecular motors and muscles. Electrochim. Acta*, 56, 10, 3721–3726, 2011.

19. Otero, T. and Cortes, M., Artificial muscle: Movement and position control. *Chem. Commun.*, 3, 284–285, 2004.

20. Barisci, J., Conn, C., Wallace, G., Conducting polymer sensors. *Trends Polym. Sci.*, 9, 4, 307–311, 1996.

21. Ramanavičius, A., Ramanavičiené, A., Malinauskas, A., Electrochemical sensors based on conducting polymer—Polypyrrole. *Electrochim. Acta*, 51, 27, 6025–6037, 2006.

22. Miasik, J.J., Hooper, A., Tofield, B.C., Conducting polymer gas sensors. *J. Chem. Soc., Faraday Trans. 1 : Physical Chemistry in Condensed Phases*, 82, 4, 1117–1126, 1986.

23. Janata, J. and Josowicz, M., Conducting polymers in electronic chemical sensors. *Nat. Mater.*, 2, 1, 19, 2003.

24. Wang, Y. and Jing, X., Intrinsically conducting polymers for electromagnetic interference shielding. *Polym. Adv. Technol.*, 16, 4, 344–351, 2005.
25. Trivedi, D.C. and Dhawan, S.K., Shielding of electromagnetic interference using polyaniline. *Synth. Met.*, 59, 2, 267–272, 1993.
26. Håkansson, E., Amiet, A., Kaynak, A., Electromagnetic shielding properties of polypyrrole/polyester composites in the 1–18 GHz frequency range. *Synth. Met.*, 156, 14-15, 917–925, 2006.
27. Dhawan, S., Singh, N., Rodrigues, D., Electromagnetic shielding behaviour of conducting polyaniline composites. *Sci. Technol. Adv. Mater.*, 4, 2, 105, 2003.
28. Rosseinsky, D.R. and Mortimer, R.J., Electrochromic systems and the prospects for devices. *Adv. Mater.*, 13, 11, 783–793, 2001.
29. Stenger-Smith, J.D., Intrinsically electrically conducting polymers. Synthesis, characterization, and their applications. *Prog. Polym. Sci.*, 23, 1, 57–79, 1998.
30. Pile, D.L. and Hillier, A.C., Electrochemically modulated transport through a conducting polymer membrane. *J. Membr. Sci.*, 208, 1-2, 119–131, 2002.
31. Ehrenbeck, C. and Jüttner, K., Development of an anion/cation permeable free-standing membrane based on electrochemical switching of polypyrrole. *Electrochim. Acta*, 41, 4, 511–518, 1996.
32. Ren, X. and Pickup, P.G., Ion transport in polypyrrole and a polypyrrole/polyanion composite. *J. Phys. Chem.*, 97, 20, 5356–5362, 1993.
33. Ariza, M.J. and Otero, T.F., Ionic diffusion across oxidized polypyrrole membranes and during oxidation of the free-standing film. *Colloids Surf., A: Physicochem. Eng. Asp.*, 270, 226–231, 2005.
34. Le, H.N.T. *et al.*, Corrosion protection and conducting polymers: Polypyrrole films on iron. *Electrochim. Acta*, 46, 26–27, 4259–4272, 2001.
35. Sitaram, S.P., Stoffer, J.O., O'Keefe, T.J., Application of conducting polymers in corrosion protection. *J. Coat. Technol.*, 69, 866, 65–69, 1997.
36. Mahmoud, W.E. *et al.*, Synthesis and characterization of electropolymerized molecularly imprinted microporous polyaniline films for solar cell applications. *Polym. Compos.*, 34, 2, 299–304, 2013.
37. Tang, Q. *et al.*, Counter electrodes from double-layered polyaniline nanostructures for dye-sensitized solar cell applications. *J. Mater. Chem. A*, 1, 2, 317–323, 2013.
38. Keothongkham, K. *et al.*, Electrochemically deposited polypyrrole for dye-sensitized solar cell counter electrodes. *Int. J. Photoenergy*, 2012, 1–7, 2012.
39. Yue, G. *et al.*, Application of poly (3, 4-ethylenedioxythiophene): Polystyrenesulfonate/polypyrrole counter electrode for dye-sensitized solar cells. *J. Phys. Chem. C*, 116, 34, 18057–18063, 2012.
40. Xia, J., Chen, L., Yanagida, S., Application of polypyrrole as a counter electrode for a dye-sensitized solar cell. *J. Mater. Chem.*, 21, 12, 4644–4649, 2011.

41. Gurunathan, K. *et al.*, Electrochemically synthesised conducting polymeric materials for applications towards technology in electronics, optoelectronics and energy storage devices. *Mater. Chem. Phys.*, 61, 3, 173–191, 1999.
42. Gao, J. *et al.*, Soluble polypyrrole as the transparent anode in polymer light-emitting diodes. *Synth. Met.*, 82, 3, 221–223, 1996.
43. Jang, J., Ha, J., Kim, K., Organic light-emitting diode with polyaniline-poly (styrene sulfonate) as a hole injection layer. *Thin Solid Films*, 516, 10, 3152–3156, 2008.
44. Yang, Y. and Heeger, A.J., Polyaniline as a transparent electrode for polymer light-emitting diodes: Lower operating voltage and higher efficiency. *Appl. Phys. Lett.*, 64, 10, 1245–1247, 1994.
45. Wang, H. *et al.*, Application of polyaniline (emeraldine base, EB) in polymer light-emitting devices. *Synth. Met.*, 78, 1, 33–37, 1996.
46. Novák, P. *et al.*, Electrochemically active polymers for rechargeable batteries. *Chem. Rev.*, 97, 1, 207–282, 1997.
47. Killian, J. *et al.*, Polypyrrole composite electrodes in an all-polymer battery system. *J. Electrochem. Soc.*, 143, 3, 936–942, 1996.
48. Osaka, T. *et al.*, Application of solid polymer electrolyte to lithium/polypyrrole secondary battery system. *J. Electrochem. Soc.*, 141, 8, 1994–1998, 1994.
49. Wang, C. *et al.*, Buckled, stretchable polypyrrole electrodes for battery applications. *Adv. Mater.*, 23, 31, 3580–3584, 2011.
50. Mermilliod, N., Tanguy, J., Petiot, F., A study of chemically synthesized polypyrrole as electrode material for battery applications. *J. Electrochem. Soc.*, 133, 6, 1073–1079, 1986.
51. Pan, L. *et al.*, Conducting polymer nanostructures: Template synthesis and applications in energy storage. *Int. J. Mol. Sci.*, 11, 7, 2636–2657, 2010.
52. Malinauskas, A., Malinauskiene, J., Ramanavičius, A., Conducting polymer-based nanostructurized materials: Electrochemical aspects. *Nanotechnology*, 16, 10, R51, 2005.
53. Ingram, M.D., Staesche, H., Ryder, K.S., 'Activated' polypyrrole electrodes for high-power supercapacitor applications. *Solid State Ionics*, 169, 1–4, 51–57, 2004.
54. Yue, B. *et al.*, Polypyrrole coated nylon lycra fabric as stretchable electrode for supercapacitor applications. *Electrochim. Acta*, 68, 18–24, 2012.
55. Jain, R., Jadon, N., Pawaiya, A., Polypyrrole based next generation electrochemical sensors and biosensors: A review. *TrAC Trends Anal. Chem.*, 97, 363–373, 2017.
56. Rahman, M. *et al.*, Electrochemical sensors based on organic conjugated polymers. *Sensors*, 8, 1, 118–141, 2008.
57. Smela, E., Conjugated polymer actuators for biomedical applications. *Adv. Mater.*, 15, 6, 481–494, 2003.
58. Brédas, J.L. and Chance, R.R., Conjugated polymeric materials: opportunities in electronics, optoelectronics, and molecular electronics, vol. 182, Springer Science & Business Media, The Netherlands, 2012.

59. Wallace, G.G. *et al.*, *Conductive electroactive polymers: intelligent materials systems*. CRC Press, Boca Raton, 2002.
60. Kaur, G. *et al.*, Electrically conductive polymers and composites for biomedical applications. *RSC Adv.*, 5, 47, 37553–37567, 2015.
61. Bakhshi, A. and Bhalla, G., Electrically conducting polymers: Materials of the twenty first century. *J. Sci. Ind. Res.*, 63, 715–728, 2004.
62. Fengel, C.V. *et al.*, Biocompatible silk-conducting polymer composite trilayer actuators. *Smart Mater. Struct.*, 26, 5, 055004, 2017.
63. Smela, E., Conjugated polymer actuators. *MRS Bull.*, 33, 3, 197–204, 2008.
64. Macdiarmid, A. *et al.*, Electrically Conducting Covalent Polymers-Halogen Derivatives of (Sn) X and (Ch) X, in: *Journal of the Electrochemical Society*, Electrochemical Soc Inc, 10 South Main Street, Pennington, NJ 08534, 1977.
65. Chiang, C.K. *et al.*, Electrical conductivity in doped polyacetylene. *Phys. Rev. Lett.*, 39, 17, 1098, 1977.
66. Basescu, N. *et al.*, High electrical conductivity in doped polyacetylene. *Nature*, 327, 6121, 403, 1987.
67. Otero, T.F., Sanchez, J.J., Martinez, J.G., Biomimetic dual sensing-actuators based on conducting polymers. Galvanostatic theoretical model for actuators sensing temperature. *J. Phys. Chem. B*, 116, 17, 5279–5290, 2012.
68. Otero, T.F., Soft, wet, and reactive polymers. Sensing artificial muscles and conformational energy. *J. Mater. Chem.*, 19, 6, 681–689, 2009.
69. Arias-Pardilla, J. *et al.*, Biomimetic Sensing–Actuators Based on Conducting Polymers, in: *Aspects on Fundaments and Applications of Conducting Polymers*, Intechopen, Croatia, 2012.
70. Otero, T. *et al.*, Electrochemomechanical properties from a bilayer: Polypyrrole/non-conducting and flexible material—Artificial muscle. *J. Electroanal. Chem.*, 341, 1-2, 369–375, 1992.
71. Otero, T. and Sansinena, J., Bilayer dimensions and movement in artificial muscles. *Bioelectrochem. Bioenerg.*, 42, 2, 117–122, 1997.
72. Otero, T.F. and Sansieña, J.M., Soft and wet conducting polymers for artificial muscles. *Adv. Mater.*, 10, 6, 491–494, 1998.
73. Huang, W.-S., Humphrey, B.D., MacDiarmid, A.G., Polyaniline, a novel conducting polymer. Morphology and chemistry of its oxidation and reduction in aqueous electrolytes. *J. Chem. Soc., Faraday Trans. 1 : Physical Chemistry in Condensed Phases*, 82, 8, 2385–2400, 1986.
74. Tsai, E. *et al.*, Anion-exchange behavior of polypyrrole membranes. *J. Phys. Chem.*, 92, 12, 3560–3565, 1988.
75. Ahonen, H.J., Lukkari, J., Kankare, J., n-and p-doped poly (3, 4-ethylenedioxythiophene): Two electronically conducting states of the polymer. *Macromolecules*, 33, 18, 6787–6793, 2000.
76. Skompska, M. *et al.*, *In situ* conductance studies of p-and n-doping of poly (3, 4-dialkoxythiophenes). *J. Electroanal. Chem.*, 577, 1, 9–17, 2005.

77. Arbizzani, C. *et al.*, N-and P-doped Polydithieno [3, 4-B: 3′, 4′-D] thiophene: A narrow band gap polymer for redox supercapacitors. *Electrochim. Acta*, 40, 12, 1871–1876, 1995.

78. Okuzaki, H., *Soft Actuators*, Springer Japan, Tokyo, 2014.

79. T.F. Otero *et al.*, Dispositivos laminares que emplean polímeros conductores capaces de provocar movimientos mecánicos. Spanish patent, 1, p. 17, 1992.

80. T. Otero, J. Rodriguez, C. Santamaria, Músculos artificiales formados por multicapas: Polímeros conductores-no conductores. Spanish patent, 12, p. 28, 1992.

81. Park, Y., Jung, J., Chang, M., Research Progress on Conducting Polymer-Based Biomedical Applications. *Appl. Sci.*, 9, 6, 1070, 2019.

82. John, S., Alici, G., Cook, C., Frequency response of polypyrrole trilayer actuator displacement, in: *Electroactive Polymer Actuators and Devices (EAPAD) 2008*, International Society for Optics and Photonics, California, United States, 2008.

83. Yao, Q., Alici, G., Spinks, G.M., Feedback control of tri-layer polymer actuators to improve their positioning ability and speed of response. *Sens. Actuators A: Phys.*, 144, 1, 176–184, 2008.

84. García-Córdova, F. *et al.*, Biomimetic polypyrrole based all three-in-one triple layer sensing actuators exchanging cations. *J. Mater. Chem.*, 21, 43, 17265–17272, 2011.

85. Della Santa, A., De Rossi, D., Mazzoldi, A., Performance and work capacity of a polypyrrole conducting polymer linear actuator. *Synth. Met.*, 90, 2, 93–100, 1997.

86. Machado, J., Karasz, F.E., Lenz, R., Electrically conducting polymer blends. *Polymer*, 29, 8, 1412–1417, 1988.

87. Park, D.S., Shim, Y.B., Park, S.M., Degradation kinetics of polypyrrole films. *J. Electrochem. Soc.*, 140, 10, 2749–2752, 1993.

88. Omastová, M. *et al.*, Chemical preparation and characterization of conductive poly (methyl methacrylate)/polypyrrole composites. *Polymer*, 39, 25, 6559–6566, 1998.

89. Migahed, M. *et al.*, Preparation, characterization, and electrical conductivity of polypyrrole composite films. *Polym. Test.*, 23, 3, 361–365, 2004.

90. Papathanassiou, A. *et al.*, Effect of hydrostatic pressure on the dc conductivity of fresh and thermally aged polypyrrole-polyaniline conductive blends. *J. Phys. D: Appl. Phys.*, 35, 17, L85, 2002.

91. Bassil, M., Davenas, J., Tahchi, M.E., Electrochemical properties and actuation mechanisms of polyacrylamide hydrogel for artificial muscle application. *Sens. Actuators B: Chem.*, 134, 2, 496–501, 2008.

92. Varghese, J.M. *et al.*, Thermoresponsive hydrogels based on poly (N-isopropylacrylamide)/chondroitin sulfate. *Sens. Actuators B: Chem.*, 135, 1, 336–341, 2008.

93. Wei, D. *et al.*, Controlled growth of polypyrrole hydrogels. *Soft Matter*, 9, 10, 2832–2836, 2013.

94. Antonio, J.L. *et al.*, Electrocontrolled Swelling and Water Uptake of a Three-Dimensional Conducting Polypyrrole Hydrogel. *ChemElectroChem*, 3, 12, 2146–2152, 2016.

95. Gilmore, K. *et al.*, Preparation of hydrogel/conducting polymer composites. *Polym. Gels Networks*, 2, 2, 135–143, 1994.

96. Kim, S.J. *et al.*, Self-oscillatory actuation at constant DC voltage with pH-sensitive chitosan/polyaniline hydrogel blend. *Chem. Mater.*, 18, 24, 5805–5809, 2006.

97. Spinks, G.M. *et al.*, A novel "dual mode" actuation in chitosan/polyaniline/carbon nanotube fibers. *Sens. Actuators B: Chem.*, 121, 2, 616–621, 2007.

98. Ismail, Y.A. *et al.*, Electrochemical actuation in chitosan/polyaniline microfibers for artificial muscles fabricated using an *in situ* polymerization. *Sens. Actuators B: Chem.*, 129, 2, 834–840, 2008.

99. Ismail, Y.A., Shin, M.K., Kim, S.J., A nanofibrous hydrogel templated electrochemical actuator: From single mat to a rolled-up structure. *Sens. Actuators B: Chem.*, 136, 2, 438–443, 2009.

100. Kim, H.I., Park, S.J., Kim, S.J., Volume behavior of interpenetrating polymer network hydrogels composed of polyacrylic acid-co-poly (vinyl sulfonic acid)/polyaniline as an actuator. *Smart Mater. Struct.*, 15, 6, 1882, 2006.

101. Gao, H. *et al.*, Monolithic polyaniline/polyvinyl alcohol nanocomposite actuators with tunable stimuli-responsive properties. *Sens. Actuators B: Chem.*, 145, 2, 839–846, 2010.

102. Shi, X. *et al.*, Electromechanical polyaniline–cellulose hydrogels with high compressive strength. *Soft Matter*, 9, 42, 10129–10134, 2013.

103. Hong, C.-H. *et al.*, Electroactive bio-composite actuators based on cellulose acetate nanofibers with specially chopped polyaniline nanoparticles through electrospinning. *Compos. Sci. Technol.*, 87, 135–141, 2013.

104. da Silva, L.B.J. and Oréfice, R.L., Synthesis and electromechanical actuation of a temperature, pH, and electrically responsive hydrogel. *J. Polym. Res.*, 21, 6, 466, 2014.

105. Bai, Y., Zhang, J., Chen, X., A thermal-, water-, and near-infrared light-induced shape memory composite based on polyvinyl alcohol and polyaniline fibers. *ACS Appl. Mater. Interfaces*, 10, 16, 14017–14025, 2018.

106. Marcasuzaa, P. *et al.*, Chitosan-graft-polyaniline-based hydrogels: Elaboration and properties. *Biomacromolecules*, 11, 6, 1684–1691, 2010.

107. Moschou, E.A. *et al.*, Voltage-switchable artificial muscles actuating at near neutral pH. *Sens. Actuators B: Chem.*, 115, 1, 379–383, 2006.

108. Lee, S.H. *et al.*, Enhanced actuation of PPy/CNT hybrid fibers using porous structured DNA hydrogel. *Sens. Actuators B: Chem.*, 145, 1, 89–92, 2010.

109. Ismail, Y.A. *et al.*, Sensing characteristics of a conducting polymer/hydrogel hybrid microfiber artificial muscle. *Sens. Actuators B: Chem.*, 160, 1, 1180–1190, 2011.

110. Wang, Y. *et al.*, Hydrogel-reinforced polypyrrole electroactuator, in: *2016 38th Annual International Conference of the IEEE Engineering in Medicine and Biology Society (EMBC)*, IEEE, Orlando, FL, USA, 2016.

111. Yamauchi, T. *et al.*, Preparation of composite materials of polypyrrole and electroactive polymer gel using for actuating system. *Synth. Met.*, 152, 1-3, 45–48, 2005.

112. Mahloniya, R., Bajpai, J., Bajpai, A., Electrical actuation of ionic hydrogels based on polyvinyl alcohol grafted with poly (2-acrylamido-2-methyl-1-propanesulfonic acid-co-acrylonitrile) chains. *Polym. Compos.*, 33, 1, 129–137, 2012.

113. Pattavarakorn, D. *et al.*, Electroactive performances of conductive polythiophene/hydrogel hybrid artificial muscle. *Energy Procedia*, 34, 673–681, 2013.

114. Pattavarakorn, D. *et al.*, Conductive polythiophene/polymer composites as electroactive material application, 2011.

115. Simaite, A. *et al.*, Hybrid PVDF/PVDF-graft-PEGMA membranes for improved interface strength and lifetime of PEDOT: PSS/PVDF/ionic liquid actuators. *ACS Appl. Mater. Interfaces*, 7, 36, 19966–19977, 2015.

116. Sekine, S. *et al.*, Conducting polymer electrodes printed on hydrogel. *J. Am. Chem. Soc.*, 132, 38, 13174–13175, 2010.

117. Dai, T.Y. et al., High-strength multifunctional conducting polymer hydrogels, in: *Advanced Materials Research*, Trans Tech Publ, Switzerland, 2010.

118. Calvo, P.A. *et al.*, Chemical oxidative polymerization of pyrrole in the presence of m-hydroxybenzoic acid- and m-hydroxycinnamic acid-related compounds. *Synth. Met.*, 126, 1, 111–116, 2002.

119. Cao, Y. *et al.*, Influence of chemical polymerization conditions on the properties of polyaniline. *Polymer*, 30, 12, 2305–2311, 1989.

120. Kudoh, Y., Akami, K., Matsuya, Y., Chemical polymerization of 3,4-ethylenedioxythiophene using an aqueous medium containing an anionic surfactant. *Synth. Met.*, 98, 1, 65–70, 1998.

121. Pron, A. *et al.*, The effect of the oxidation conditions on the chemical polymerization of polyaniline. *Synth. Met.*, 24, 3, 193–201, 1988.

122. Bajpai, A., Bajpai, J., Soni, S., Preparation and characterization of electrically conductive composites of poly (vinyl alcohol)-g-poly (acrylic acid) hydrogels impregnated with polyaniline (PANI). *Express Polym. Lett.*, 2, 26–39, 2008.

123. Sharma, K. *et al.*, Application of biodegradable superabsorbent hydrogel composite based on Gum ghatti-co-poly (acrylic acid-aniline) for controlled drug delivery. *Polym. Degrad. Stab.*, 124, 101–111, 2016.

124. Molina, M., Rivarola, C., Barbero, C., Effect of copolymerization and semi-interpenetration with conducting polyanilines on the physicochemical properties of poly (N-isopropylacrylamide) based thermosensitive hydrogels. *Eur. Polym. J.*, 47, 10, 1977–1984, 2011.

125. Xia, Y. and Zhu, H., Polyaniline nanofiber-reinforced conducting hydrogel with unique pH-sensitivity. *Soft Matter*, 7, 19, 9388-9393, 2011.

126. Adhikari, S. and Banerji, P., Polyaniline composite by in situ polymerization on a swollen PVA gel. *Synth. Met.*, 159, 23-24, 2519-2524, 2009.

127. Gniadek, M. *et al.*, Construction of multifunctional materials by intrachannel modification of NIPA hydrogel with PANI-metal composites. *J. Electroanal. Chem.*, 812, 273-281, 2018.

128. Hao, G.-P. *et al.*, Stretchable and semitransparent conductive hybrid hydrogels for flexible supercapacitors. *ACS Nano*, 8, 7, 7138-7146, 2014.

129. Lee, H.-J. *et al.*, Fabrication and evaluation of bacterial cellulose-polyaniline composites by interfacial polymerization. *Cellulose*, 19, 4, 1251-1258, 2012.

130. Dai, T. *et al.*, Interfacial polymerization to high-quality polyacrylamide/polyaniline composite hydrogels. *Compos. Sci. Technol.*, 70, 3, 498-503, 2010.

131. Thanpitcha, T. *et al.*, Preparation and characterization of polyaniline/chitosan blend film. *Carbohydr. Polym.*, 64, 4, 560-568, 2006.

132. Kim, S.J. *et al.*, Synthesis and characteristics of a semi-interpenetrating polymer network based on chitosan/polyaniline under different pH conditions. *J. Appl. Polym. Sci.*, 96, 3, 867-873, 2005.

133. Ogura, K. *et al.*, The humidity dependence of the electrical conductivity of a solublepolyaniline–poly (vinyl alcohol) composite film. *J. Mater. Chem.*, 7, 12, 2363-2366, 1997.

134. Dmitriev, I.Y. *et al.*, Mechanical response and network characterization of conductive polyaniline/polyacrylamide gels. *Mater. Chem. Phys.*, 187, 88–95, 2017.

135. Li, L. *et al.*, In situ forming biodegradable electroactive hydrogels. *Polym. Chem.*, 5, 8, 2880-2890, 2014.

136. Huang, H. *et al.*, Reinforced polyaniline/polyvinyl alcohol conducting hydrogel from a freezing–thawing method as self-supported electrode for supercapacitors. *J. Mater. Sci.*, 51, 18, 8728-8736, 2016.

137. Tang, Q. *et al.*, Polyaniline/polyacrylamide conducting composite hydrogel with a porous structure. *Carbohydr. Polym.*, 74, 2, 215-219, 2008.

138. Dou, P. *et al.*, Rapid synthesis of hierarchical nanostructured Polyaniline hydrogel for high power density energy storage application and three-dimensional multilayers printing. *J. Mater. Sci.*, 51, 9, 4274-4282, 2016.

139. Jiang, W. *et al.*, Separation-Free Polyaniline/TiO2 3D Hydrogel with High Photocatalytic Activity. *Adv. Mater. Interfaces*, 3, 3, 1500502, 2016.

140. Pan, L. *et al.*, Hierarchical nanostructured conducting polymer hydrogel with high electrochemical activity. *Proc. Natl. Acad. Sci.*, 109, 24, 9287-9292, 2012.

141. Sun, K.-H. *et al.*, Evaluation of in vitro and in vivo biocompatibility of a myo-inositol hexakisphosphate gelated polyaniline hydrogel in a rat model. *Sci. Rep.*, 6, 23931, 2016.

142. Wang, K. *et al.*, Flexible solid-state supercapacitors based on a conducting polymer hydrogel with enhanced electrochemical performance. *J. Mater. Chem. A*, 2, 46, 19726–19732, 2014.

143. Wu, H. *et al.*, Stable Li-ion battery anodes by *in-situ* polymerization of conducting hydrogel to conformally coat silicon nanoparticles. *Nat. Commun.*, 4, 1943, 2013.

144. Yan, B. *et al.*, Fabrication of polyaniline hydrogel: Synthesis, characterization and adsorption of methylene blue. *Appl. Surf. Sci.*, 356, 39–47, 2015.

145. Xu, G. *et al.*, Porous nitrogen and phosphorus co-doped carbon nanofiber networks for high performance electrical double layer capacitors. *J. Mater. Chem. A*, 3, 46, 23268–23273, 2015.

146. Zhai, D. *et al.*, Highly sensitive glucose sensor based on Pt nanoparticle/polyaniline hydrogel heterostructures. *ACS Nano*, 7, 4, 3540–3546, 2013.

147. Heydari, H. and Gholivand, M.B., An all-solid-state asymmetric device based on a polyaniline hydrogel for a high energy flexible supercapacitor. *New J. Chem.*, 41, 1, 237–244, 2017.

148. Das, S. *et al.*, Enhancement of energy storage and photoresponse properties of folic acid–polyaniline hybrid hydrogel by in situ growth of Ag nanoparticles. *ACS Appl. Mater. Interfaces*, 8, 41, 28055–28067, 2016.

149. Wang, Y. *et al.*, Dopant-enabled supramolecular approach for controlled synthesis of nanostructured conductive polymer hydrogels. *Nano Lett.*, 15, 11, 7736–7741, 2015.

150. Pan, L. *et al.*, An ultra-sensitive resistive pressure sensor based on hollow-sphere microstructure induced elasticity in conducting polymer film. *Nat. Commun.*, 5, 3002, 2014.

151. Shi, Y. *et al.*, Nanostructured conductive polypyrrole hydrogels as high-performance, flexible supercapacitor electrodes. *J. Mater. Chem. A*, 2, 17, 6086–6091, 2014.

152. Park, S.-B. *et al.*, A novel pH-sensitive membrane from chitosan—TEOS IPN; Preparation and its drug permeation characteristics. *Biomaterials*, 22, 4, 323–330, 2001.

153. Smela, E., Lu, W., Mattes, B.R., Polyaniline actuators: Part 1. PANI (AMPS) in hcl. *Synth. Met.*, 151, 1, 25–42, 2005.

154. Lewis, T.W. *et al.*, Evaluation of solid polymer electrolytes for use in conducting polymer/nanotube actuators, in: *Smart Structures and Materials 2000: Electroactive Polymer Actuators and Devices (EAPAD)*, International Society for Optics and Photonics, Newport Beach, CA, United States, 2000.

155. Flory, P.J., *Principles of polymer chemistry*. Cornell University Press, Ithaca, New York, 1953.

156. Tanaka, T. *et al.*, Collapse of gels in an electric field. *Science*, 218, 4571, 467–469, 1982.

157. Ma, Y. *et al.*, *In situ* fabrication of a water-soluble, self-doped polyaniline nanocomposite: The unique role of DNA functionalized single-walled carbon nanotubes. *J. Am. Chem. Soc.*, 128, 37, 12064–12065, 2006.

158. Zanuy, D. and Alemán, C., DNA– Conducting Polymer Complexes: A Computational Study of the Hydrogen Bond between Building Blocks. *J. Phys. Chem. B*, 112, 10, 3222–3230, 2008.

159. Chen, G.Z. *et al.*, Carbon nanotube and polypyrrole composites: Coating and doping. *Adv. Mater.*, 12, 7, 522–526, 2000.

160. Spinks, G.M. *et al.*, Conducting polymer, carbon nanotube, and hybrid actuator materials, in: *Smart Structures and Materials 2001: Electroactive Polymer Actuators and Devices*, International Society for Optics and Photonics, Newport Beach, CA, United States, 2001.

161. Ryu, K.S. *et al.*, The electrochemical performance of polythiophene synthesized by chemical method as the polymer battery electrode. *Mater. Chem. Phys.*, 84, 2-3, 380–384, 2004.

162. Bangyekan, C., Aht-Ong, D., Srikulkit, K., Preparation and properties evaluation of chitosan-coated cassava starch films. *Carbohydr. Polym.*, 63, 1, 61–71, 2006.

163. Hiamtup, P., Sirivat, A., Jamieson, A.M., Electromechanical response of a soft and flexible actuator based on polyaniline particles embedded in a cross-linked poly (dimethyl siloxane) network. *Mater. Sci. Eng. C*, 28, 7, 1044–1051, 2008.

164. Ido, Y. *et al.*, Conducting polymer microelectrodes anchored to hydrogel films. *ACS Macro Lett.*, 1, 3, 400–403, 2012.

165. Inganas, O. *et al.*, Model polymers for polymer actuators, in: *Smart Structures and Materials 1999: Electroactive Polymer Actuators and Devices*, International Society for Optics and Photonics, Newport Beach, CA, United States, 1999.

166. Bahadur, P. and Sastry, N., *Principles of polymer science*, Alpha Science Int'l Ltd, Oxford, UK, 2005.

167. Rudin, A. and Choi, P., *The elements of polymer science and engineering*, Academic Press, UK, 2012

Index

Also of Interest

Check out these other forthcoming and published titles from Scrivener Publishing

By the same editors:

Applications for Metal-Organic Frameworks and Their Derived Materials, edited by Inamuddin, Rajender Boddula, Mohd Imran Ahamed, and Abdullah M. Asiri, ISBN 9781119650980. Edited by one of the most well-respected and prolific engineers in the world and his team, this is the most thorough, up-to-date, and comprehensive volume on metal-organic frameworks and their derived materials available today. *COMING IN MAY 2020*

Potassium-Ion Batteries: Materials and Applications, edited by Inamuddin, Rajender Boddula, and Abdullah M. Asiri, ISBN 9781119661399. Edited by one of the most well-respected and prolific engineers in the world and his team, this is the most thorough, up-to-date, and comprehensive volume on potassium-ion batteries available today. *COMING IN APRIL 2020*

Zinc Batteries: Basics, Developments, and Applications , edited by Rajender Boddula, Inamuddin, and Abdullah M. Asiri, ISBN 9781119661894. Edited by one of the most well-respected and prolific engineers in the world and his team, this is the most thorough, up-to-date, and comprehensive volume on zinc batteries available today. *COMING IN APRIL 2020*

Rechargeable Batteries: History, Progress, and Applications, edited by Rajender Boddula, Inamuddin, Ramyakrishna Pothu, and Abdullah M. Asiri, ISBN 9781119661191. Edited by one of the most well-respected and prolific engineers in the world and his team, this is the most thorough, up-to-date, and comprehensive volume on rechargeable batteries available today. *COMING IN APRIL 2020*

Other Books of Interest:

Energy Storage: A New Approach 2nd Edition, by Ralph Zito and Haleh Ardebili, ISBN 9781119083597. A revision of the groundbreaking study of methods for storing energy on a massive scale to be used in wind, solar, and other renewable energy systems. *NOW AVAILABLE!*

Operator's Guide to Process Compressors, by Robert X. Perez, ISBN 9781119580614. The perfect primer for anyone responsible for operating or maintaining process gas compressors. *NOW AVAILABLE!*

Troubleshooting Rotating Machinery: Including Centrifugal Pumps and Compressors, Reciprocating Pumps and Compressors, Fans, Steam Turbines, Electric Motors, and More, by Robert X. Perez and Andrew P. Conkey ISBN 9781119294139. A must-read for anyone whose job depends on operating or maintaining safe, reliable, and efficient machinery in a process facility, this volume shows the newcomer and veteran alike how you can start troubleshooting rotating equipment problems immediately by implementing this book's common sense methodology. *NOW AVAILABLE!*

Operator's Guide to General Purpose Steam Turbines: An Overview of Operating Principles, Construction, Best Practices, and Troubleshooting, by Robert X. Perez and David W. Lawhon, ISBN 9781119294214. A must-read book for anyone operating or maintaining general purpose steam turbines in a process setting, this is a practical reference that includes a collection of field-tested procedures aimed at ensuring safe general purpose steam turbines start-ups, shutdowns, and over-speed tests. *NOW AVAILABLE!*

CPSIA information can be obtained
at www.ICGtesting.com
Printed in the USA
LVHW080134100620
656365LV00005B/19